Biometrics

Challenges, Trends and Opportunities

Editors

Francisco Liébana-Cabanillas
Faculty of Economics and Business Administration
University of Granada, Spain

Francisco Muñoz-Leiva
Faculty of Economics and Business Administration
University of Granada, Spain

Juan F. Prados-Castillo
Faculty of Social and Legal Sciences of Melilla
University of Granada, Spain

Elena Higueras-Castillo
Faculty of Economics and Business Administration
University of Granada, Spain

CRC Press
Taylor & Francis Group
Boca Raton London New York

CRC Press is an imprint of the
Taylor & Francis Group, an **informa** business
A SCIENCE PUBLISHERS BOOK

First edition published 2026
by CRC Press
2385 NW Executive Center Drive, Suite 320, Boca Raton FL 33431

and by CRC Press
4 Park Square, Milton Park, Abingdon, Oxon, OX14 4RN

CRC Press is an imprint of Taylor & Francis Group, LLC

Library of Congress Cataloging-in-Publication Data (applied for)

ISBN: 978-1-032-81430-8 (hbk)
ISBN: 978-1-032-83489-4 (pbk)
ISBN: 978-1-003-50955-4 (ebk)

DOI: 10.1201/9781003509554

Typeset in Times New Roman
by Prime Publishing Services

Acknowledgements

We would like to express our deepest gratitude to all the authors who contributed to the creation of this book, and the reviewers who helped us improve the initial documents. Finally, we would like to thank the researchers, educators, and the scientific community for generating this book in the innovative field of biometrics. This book is a collective effort of all the editors and authors referenced in it.

Preface

In the current era of digital transformation, biometrics technologies have emerged as essential tools for tasks, such as identity verification, offering innovative solutions that balance security and convenience. Their applications range from authentication on financial services platforms to access to healthcare systems or citizen databases, redefining the interaction between humans and technology, and facilitating faster and more efficient access to digital services.

This book examines the impact of such biometrics technologies from a multidisciplinary perspective, addressing their integration with other technologies, such as artificial intelligence, machine learning, and blockchain. It presents cases in sectors such as banking, health, and tourism, paying special attention to opportunities, challenges, and risks related to security concerns, compliance with different data protection regulations, and implementation costs. Throughout the chapters, the book provides an in-depth analysis of emerging trends in the field of biometrics, exploring innovations such as multi-modal authentication, biometric payment systems, and their intersection with regulation and ethics. Improvements in accuracy and adaptability driven by the combination with other technologies, such as machine learning, are highlighted, as well as the challenges associated with the variability of biometric data in real-world conditions.

Owing to the high transformative potential of these technologies, it is essential to examine their impact from a critical but constructive perspective. This book aims to provide a balanced view between theoretical frameworks and practical applications, serving as a key reference for researchers, practitioners, and decision-makers interested in the evolution of biometrics and its role in the digital economy.

We would like to thank the authors for their work, which allowed us to consolidate this work. The editors hope that this book will contribute to a better understanding of biometrics and ideas for future research.

Francisco Liébana-Cabanillas
Francisco Muñoz-Leiva
Juan F. Prados-Castillo
Elena Higueras-Castillo

About the Contributors

Francisco Liébana-Cabanillas holds the position of Full Professor at the University of Granada, Spain, with expertise in mobile and online banking, Internet consumer behavior, and digital marketing. He is included in Stanford University's top 2% of scientists' lists in 2021– 2024.

Francisco Muñoz-Leiva holds the position of Full Professor at the University of Granada, Spain, with expertise in Internet consumer behavior, mobile and online banking, research methodologies, and consumer neuroscience. His significant scholarly contributions have resulted in his inclusion in Stanford University's top 2% of scientists listed for 2022–2024.

Juan F. Prados-Castillo is a Permanent Lecturer in the Department of Marketing and Market Research at the University of Granada, Spain. His research and teaching endeavors concentrate on tourism, the digital economy, and emerging technologies, with the objective of enhancing business efficiency and evaluating consumer impact.

Elena Higueras-Castillo is an Assistant Lecturer in the Department of Marketing and Market Research at the University of Granada, Spain. Her area of specialization is digital marketing. Her primary research focuses on the adoption of environmentally friendly technologies and mobile and online banking services.

Aayushi Kalra, an Assistant Professor specializing in business analytics, with expertise in machine learning, data visualization, and supply chain analytics, enriched by her academic and practical contributions to the field.

Juan Antonio Torrecilla-García is based at the Economics and Business Administration Department, University of Málaga. He completed his PhD in Business Administration and Human Resources Management, degree in economics and business studies, degree in labor sciences, diploma in business studies, and MBA in business management and administration.

Agnieszka Grazyna Skotnicka, an innovation consultant specializing in innovation planning, business opportunity diagnosis, technological and social entrepreneurship, innovative product marketing, and change management projects. With 25 years of experience in Spain's consulting sector, she served as the Consulting Director for a few different companies.

Alvaro Saavedra holds a PhD in Economics, Business, and Law from the Public University of Navarra. His research focuses on human-computer interaction, specifically on consumer-business interactions through AI applications such as virtual assistants, employing consumer neuroscience methodologies and experimental approaches.

Ana Moreno-Lobato holds a PhD in Tourism from the Universidad de Extremadura. Her research focuses on tourism marketing, specifically on consumer behavior, sensory marketing, and gastro tourism, analyzing tourist reactions to sensory tools through consumer neuroscience methodologies and experimental approaches. She currently teaches marketing at the Universidad de Extremadura, Spain.

Aroa Costa-Feito specializes in consumer neuroscience, bringing 10 years of experience leading marketing, communication, and graphic design departments in Spanish and international companies. She currently contributes to European projects and teaches at the University of León as part of its research team.

Zoran Kalinic, a Professor at the Faculty of Economics, University of Kragujevac, Serbia, holds a PhD degree in Software Engineering from the University of Kragujevac. His current research interests include digital transformation, e-business, mobile and e-commerce, payment systems, and the application of artificial intelligence in business and management.

Jana Prodanova, a marketing Professor whose research specializes in human-computer interaction and cognitive science, with a focus on usability, user experience, and the adoption of emerging technologies in digital interfaces.

Anita Ciunova-Shuleska, a Professor specializing in marketing strategies and consumer behavior, focusing on services, new technologies, and firm performance.

Nikolina Palamidovska-Sterjadovska, a Professor specializing in consumer behavior, customer engagement, and marketing communication, focusing on the adoption of new technologies, such as mobile banking and social media.

Ljupco Kocarev, a computer science Professor whose social science research focuses on human behavior, societal impacts of technology, and interdisciplinary approaches to understanding social dynamics in the digital age.

Justyna Ziobrowska-Sztuczka is Dr. in the discipline of Economics and Finance, specializing in the sharing economy and modern forms of business finance. She is the Vice President of the Polish Economic Society Wroclaw Branch, Institute of Apidology, and Lecturer at Children's University.

Dominika Łyś, a Master's student in Economics, specialization: tax consultancy; preparing her thesis entitled, *Sustainable Tourism as a Manifestation of Sharing Economy in the Face of Crises.* Academically interested in tax law and financial and tax accounting issues.

Dr. Aijaz A. Shaikh is working as Senior Research Fellow at the Institute of Information and Computational Technologies, Almaty, Kazakhstan. He holds a PhD degree in Digital Marketing from the University of Jyväskylä in Finland and professional certification in "digital analytics and digital business" from Columbia Business School, USA.

Eva R. Porras, a PhD in finance with over 20 years of experience in academic research and teaching, specializing in bubbles and contagion in financial markets, and blockchain technology.

Konstantinos Sgantzos, a Theoretical Physicist and Biochemist, with an MSc in Computational Biology & Bioinformatics. He is academically affiliated and pursuing his PhD in AI Ethics with School of Applied Mathematics and Physics, at the National Technical University of Athens, Greece.

Contents

Part II: Challenges, Comparisons, and the Future of Biometrics

Introduction

The field of biometrics has evolved significantly in recent years. Basic fingerprint recognition technology is used to develop advanced solutions capable of integrating other technologies, such as artificial intelligence (AI) and machine learning (ML), to address increasingly complex problems related to security and user experience. These technologies rely on unique physical and behavioral characteristics, such as voice and facial expressions, and redefine authentication and human-computer interaction (Kreibig, 2010; Jain et al., 2016). Biometrics offer greater accuracy and speed in identity verification but pose technical and ethical challenges related to privacy, fairness, and balance in systems (Venkatraman & Delpachitra, 2008; Daugman, 2009).

Among the emerging trends from a biometric technology perspective, the use of AI and ML has overcome the limitations of nascent biometric systems, such as reliance on predefined rules and the existence of biometric data variability issues in real-world conditions. They can handle large volumes of data and improve the accuracy and adaptability of systems to the changing characteristics of human users (Genovese et al., 2018; Mahmoud et al., 2023). In sectors such as banking, biometric payments have proven effective in preventing different types of fraud and improving customer experience. Methods such as fingerprint and voice recognition are used for security and convenience (Ahmed et al., 2021; Kambampati et al., 2024).

Despite its advantages, the mass adoption of biometrics presents significant barriers to its implementation. Key challenges include high implementation costs and a lack of interoperability between systems. Of particular note are concerns about the ethical issues surrounding the use and storage of sensitive and personal user data. Finally, public perception and user acceptance of the technology should also be considered, as this varies depending on the cultural context and local regulations that hinder its adoption as a global solution (Lewandowski, 2017; Odorović et al., 2020). These limitations call for the development of more inclusive standards and actions to strengthen regulations and ensure a balance between security and privacy (Regulation - 2016/679 - EN - GDPR-EUR-Lex, 2016).

This book, *Biometrics: Challenges, Trends, and Opportunities*, discusses the transformative potential of biometric technologies to address challenges related to areas such as security, user experience, and digital identity. By integrating insights from diverse areas and technologies, such as AI , ML , behavioral science, and data protection, this book presents a comprehensive view of the current landscape of biometric technology and its evolving role in various industries and economic areas. Its chapters aim to provide a balance between theoretical frameworks and practical applications. These issues have raised and explored topics such as multimodal authentication, biometric payment systems, and the integration of biometrics with blockchain technology. Emphasis is placed on the ethical considerations, regulatory frameworks, and technological innovations that shape the adoption of biometrics.

To improve the structure and organization of the book and to consider the objectives and scope of the carefully selected chapters included in this book, it has been organized into two sections. The first part deals with the fundamentals and applications of biometric technology. The second section focuses on the challenges, comparisons, and future of biometric technologies. A brief summary of each chapter is as follows.

Chapter 1 examines how emerging technologies such as ML and AI are transforming the performance of biometric systems, with advancements in facial, voice, fingerprint, and behavioral recognition. The transition from traditional biometrics to AI-enhanced solutions raises concerns regarding accuracy, bias, privacy, and regulatory compliance when offering applications in various industries. Future trends, such as integration with upcoming technologies and behavior-based recognition, are also highlighted. AI and ML will revolutionize identity verification, enhance security, and shape the future of biometric authentication through data analysis.

Chapter 2 explores how biometrics—measurements of physiological and behavioral traits—offer objective, real-time insights into users' cognitive and emotional states, addressing the limitations of traditional (user experience) UX methods, such as self-reports and observational data. Through theoretical analysis, scoping reviews, and case studies, this study underscores how biometric techniques, such as eye tracking, facial expression analysis, and electroencephalography (EEG), enable detailed, continuous UX data collection. They reveal subconscious reactions and microexpressions, aiding the design of personalized and adaptive user interfaces, particularly in augmented reality (AR) and AI systems. Case studies have demonstrated the transformative use of biometrics in healthcare, hospitality, and gaming.

Chapter 3 explores the theoretical knowledge and applications of three biometric tools: face reading, eye tracking, and electrodermal activity. The versatility, usability, application areas, and development options of each tool were examined. The results highlight new potential trends and paradigms in consumer research using implicit metrics. The conclusion emphasizes that these tools enhance the understanding of consumer behavior, provided that experimental protocols are correctly executed while considering limitations and biases.

Chapter 4 examines the transformative impact of biometric authentication systems on banking security and customer experience. It analyzes biometric technologies, such as facial recognition, fingerprint scanning, palm vein recognition, voice identification, and iris recognition, demonstrating their effectiveness in reducing fraud, phishing, and spoofing. Case studies from BBVA, EVO Banco, Seven Bank, Banco Bradesco, and QNB illustrate how biometrics enhance operational efficiency, regulatory compliance, and customer satisfaction. Despite these benefits, challenges such as high implementation costs, technological barriers, and data privacy concerns remain. Future research should focus on multimodal biometric solutions, improving interoperability standards, and ensuring privacy-by-design approaches.

Chapter 5 examines the integration potential and challenges of biometrics in Macedonian banks using qualitative analysis based on interviews with bank managers and IT professionals. The interviews assessed the convenience, operational impact, consumer experience, and security risks of biometrics. Insights from banking and IT experts highlight the benefits and drawbacks of implementing biometrics. Data were analyzed using thematic, sentiment, and content analyses to identify recurring themes, word frequencies, key insights, and the emotional tone of the interviews. The findings offer a comprehensive understanding of the current state, benefits, and challenges of biometric technology in Macedonian banking, providing insights for future strategies to enhance security, regulations, efficiency, and user experience in financial services.

Chapter 6 analyzes Poles' acceptance of and willingness to use biometric payments and identifies potential barriers to their adoption. Literature analysis, critiques, diagnostic surveys, and basic statistical analyses were conducted. A pilot study of 377 Poles found those aged

27–36 and 37–50 were most receptive to biometric payments, while those under 18 were the least receptive. The survey highlighted the key advantages of biometric payments, including no need for a smartphone or wallet, faster transaction speeds, high hygiene, and convenience. This study suggests a rising demand for modern financial technologies that drive the growth of Poland's digital payment sector. Technology companies and banks can invest in biometric payment infrastructure as society becomes more open to innovation.

Chapter 7 examines user adoption of fingerprint and voice recognition payment systems, extending the Unified Theory of Acceptance and Use of Technology (UTAUT2) by integrating Perceived Risk and Trust Theory. A survey of 224 potential BPS users assessed their adoption intention. The results indicated that performance expectancy and social influence positively influenced BPS adoption intentions, whereas perceived risk was insignificant. Additionally, the type of biometric system affects user behavior; social influence is more significant for voice payment adoption, whereas habit is more important for fingerprint payment adoption. These insights have theoretical and practical implications, stressing the importance of trust and social influence in promoting biometric payments. Future research should investigate other biometric technologies, such as facial recognition, and consider their longitudinal directions.

Chapter 8 examines the promising applications of this synergy, including multimodal authentication, sensitive data protection, and improved digital identity management. It also addresses challenges such as latency, implementation costs, and privacy-related legal concerns. Through a literature review and applied examples, the benefits and limitations of this technological integration were analyzed, underscoring its potential to revolutionize critical systems. This study lays the groundwork for future research and development on the application of blockchain and biometrics across various business and social contexts.

References

Ahmed, W., Rasool, A., Javed, A.R., Kumar, N., Gadekallu, T.R., Jalil, Z., & Kryvinska, N. (2021). Security in next generation mobile payment systems:

A comprehensive survey. *IEEE Access*, 9, 115932–115950. https://doi.org/10.1109/ACCESS.2021.3105450.

Daugman, J. (2009). How iris recognition works. In: *The Essential Guide to Image Processing* (pp. 715–739). Elsevier. https://doi.org/10.1016/B978-0-12-374457-9.00025-1.

Genovese, A., Muñoz, E., Piuri, V., & Scotti, F. (2018). Advanced biometric technologies: Emerging scenarios and research trends. In: P. Samarati, I. Ray, & I. Ray (Eds.), *From Database to Cyber Security* (Vol. 11170, pp. 324–352). Springer International Publishing. https://doi.org/10.1007/978-3-030-04834-1_17.

Jain, A.K., Nandakumar, K., & Ross, A. (2016). 50 years of biometric research: Accomplishments, challenges, and opportunities. *Pattern Recognition Letters*, 79, 80–105. https://doi.org/10.1016/j.patrec.2015.12.013.

Kambampati, P., Rane, S., Shoeb, A., & Dhannawat, R. (2024). PAYV – Payment Voice: A platform using voice recognition to enable payment transactions. *2024 Asia Pacific Conference on Innovation in Technology* (*APCIT*), 1– 6. https://doi.org/10.1109/APCIT62007.2024.10673442.

Kreibig, S.D. (2010). Autonomic nervous system activity in emotion: A review. *Biological Psychology*, 84(3), 394–421. https://doi.org/10.1016/j.biopsycho.2010.03.010.

Lewandowski, R. (2017). Biometria – nowe zastosowania. *Przegląd Bezpieczeństwa Wewnętrznego*, 9(17), 152–166.

Odorović, A., McKain, G., Garvey, K., Schizas, E., Zhang, B.Z., Rowan, P., & Ziegler, T. (2020). FinTech innovation in the Western Balkans: Policy and regulatory implications and potential interventions. *SSRN Electronic Journal*. https://doi.org/10.2139/ssrn.3619214.

Regulation - 2016/679 - EN - GDPR- EUR-Lex (2016). https://eur-lex.europa.eu/eli/reg/2016/679/oj.

Tareck Mahmoud, A., A. Awad, W., Behery, G., Abouhawwash, M., Masud, M., Aljuaid, H., & Ismail Ebada, A. (2023). An automatic deep neural network model for fingerprint classification. *Intelligent Automation & Soft Computing*, 36(2), 2007–2023. https://doi.org/10.32604/iasc.2023.031692.

Venkatraman, S., & Delpachitra, I. (2008). Biometrics in banking security: A case study. *Information Management & Computer Security*, 16(4), 415–430. https://doi.org/10.1108/09685220810908813.

Part I:
Fundamentals and Applications of Biometrics

1

Emerging Technologies in Biometrics: Artificial Intelligence and Machine Learning

Aayushi Kalra(0009-0007-2281-2746)
Jaipuria Institute of Management Noida, India

1. Introduction

Biometric systems have evolved from simple fingerprinting to the sophisticated Artificial Intelligent (AI) integrated systems (Modi et al., 2022; Genovese et al., 2018). Biometric identifications began with the manual matching of simple and static features like fingerprints and facial traits (Albalawi et al., 2022). These pioneering and innovative biometrics systems were inaccurate and thus had limited applicability. The advancements in technology have led to development in complexity and capabilities of biometric systems with the inclusion of several forms of identity verification, namely, iris scanning and voice recognition (Sreeja et al., 2018).

Identity verification is a dynamic process of learning and adapting to subtle and unique characteristics of individuals in fingerprints and facial features (Jain, 2015). Emerging technologies like AI and ML in biometrics have made it possible. Biometric systems have been fundamentally transformed by these technologies that allow biometrics to process vast amounts of data, identify complex patterns, and improve their accuracy over time. There is an appalling need for secure and efficient authentication methods in the age of digital financial transactions and hence, AI-driven biometrics have evolved as a key solution.

The efficiency of financial transactions along with the security have been significantly enhanced with the integration of AI and ML

Email: aayushi.kalra@jaipuria.ac.in

into biometric systems (Alzamel et al., 2019). Traditional methods of authentication, including passwords and PINs and conventional biometrics can fail in sophisticated fraud situations because of their vulnerable to breaches (Bah et al., 2019). The capability of AI and ML to continuously learn from each interaction (datapoint) and on their basis detecting anomalies and blocking fraudulent activities accurately is remarkable. This was infeasible by earlier technologies. Therefore, AI-integrated biometrics became a cornerstone for identity verification and digital security in today's interconnected world.

The chapter is structured as follows. Section 2 presents the Evolution of Biometric Technologies. Section 3 describes the fundamentals of ML and AI in biometrics. Section 4 reviews the Transformative Impact of AI and ML on Biometric Modalities. Section 5 discusses the application of AI-based Biometrics across Sectors. Section 6 presents Challenges and Ethical Considerations, and Section 7 discusses Future Trends in AI and MLe for Biometrics. Finally, Section 8 concludes the chapter.

2. The Evolution of Biometric Technologies

Biometric systems are based on the physiological and behavioral characteristics of individuals. Physiological characteristics include face, fingerprint, palm-print, and iris of a human whereas behavioral traits are voice and gait. These traits came into practice centuries ago for recognizing individuals as humans. The use of these systems was initiated in the late 19th century when Sir Francis Galton pioneered the scientific study of fingerprints. His work established the uniqueness and permanence of fingerprints providing the basis for widespread use in law enforcement and civil identification. Galton's *Finger Prints* published in 1892 became pivotal in the field of modern biometric systems.

Fingerprinting quickly became a keystone of biometric identification in criminal investigations (Dey et al., 2020). The particular patterns in minutiae points (ridges and valleys of the fingertip) are highly distinct and unique for each person and thus can differentiate two persons. Due to its simplicity and low cost, fingerprinting spread rapidly worldwide, marking the beginning of the revolution in biometrics. The emergence of other biometric modalities, like facial recognition, voice recognition, and iris scanning began thereafter.

Facial recognition system involves a simple photographic comparison from a digital image/video frame from a video source for the identification/verification of an individual (Kortli et al., 2020). It recognizes a person based on the patterns in facial textures and shapes. In the mid-1960s Woody Bledsoe, Helen Chan Wolf, and Charles Bisson pioneered the earliest automatic facial recognition systems using computers (Pampus et al., 1998). These systems applied basic algorithms to measure distances between facial features like the eyes, nose, and mouth, called landmarks. Their applicability was limited because of its performance being highly dependent upon the complexities in head inclination, face expression, and illumination (Wu et al., 2019).

In the 1980s, John Daugman developed the Iris recognition system who patented his iris recognition algorithm in 1994, which became the industry standard for iris scanning systems (Daugman, 2009). It could identify individuals based on patterns in the colored area of the eye with high accuracy and low matching times. This technology is more secure than fingerprinting, as it's difficult to replicate or alter iris than fingerprints (Daugman, 2001).

The advancements in speech processing and pattern recognition during this period led to another biometric recognition system: voice recognition. These early systems could not handle the variability of speech, including accent, tone of voice, and background noise, but the introduction of sophisticated AI algorithms enabled better voice verification (Campbell, 1997).

With the emergence of digitalization, biometrics has undergone a transformation characterized by the metamorphosis-convergence of the existing technologies, advancements in processing capabilities, means of data storage, and algorithms for pattern recognition (Goode, 2018). Multibiometric systems have evolved as the modern-day biometric systems, these fuse multiple biometric modalities, namely, fingerprint, iris, and facial recognition to improve accuracy and robustness for verifying individuals in highly secure environments. These developments presented certain challenges related to managing large databases and adapting to complex, real-world situations for traditional biometric systems.

Traditional biometric systems operate based on pre-set algorithms that are heavily dependent on static data. These use a set of rigid rules for matching and verification purposes. Fingerprint recognition technologies check for predefined features on an individual's fingerprint, the so-called minutiae, against those stored in some database. Similarly, the old face recognition systems would analyze a few of the geometric features, like the distances between the eyes, nose, and mouth, comparing those against some pre-set template. In ideal conditions, it worked quite well;

traditionally, it had been hard to implement in real life due to variables in illumination, pose, and background noise that led to discrepancies and inaccuracies.

A major limitation of the traditional systems was that they could not handle temporal variations: face recognition systems might be affected by aging, weight increase/decrease, or some type of changes in face structure, while fingerprint recognition might be affected by skin cuts or injuries (Bhatt et al., 2010). Moreover, traditional systems often lacked the ability to handle large, diverse datasets, making them vulnerable to errors and false positives when used in complex real-life situations.

Contrastingly, AI-integrated systems represent a significant advancement in biometric technology. They are developed using ML and AI models, which learn from vast datasets and have the ability to improve automatically over time. Moreover, AI modes, specifically deep learning (DL) models, are not limited to predefined rules. Traditional algorithms have limitations in analyzing complex data patterns, whereas ML algorithms work with the data with higher precision and accuracy.. AI-driven biometric systems also account for variations in lighting, pose, and physiological changes over time because of their adaptability towards the new data.

Facial recognition systems can determine important features from an image with the Convolutional neural networks (CNNs), a special type of DL architecture. The variations in different poses and lighting conditions are adjusted in the CNN model. These systems have outperformed traditional algorithms in terms of accuracy and robustness. Compared to traditional biometric systems, facial recognition systems using deep neural networks, perform with higher accuracy (Muhammad et al., 2021).

Similarly, ML models used in AI-driven fingerprinting recognition systems recognize fingerprints with a higher accuracy (Mahmoud et al., 2023). Traditional systems are static, less accurate. and produce errors because of temporary changes in fingerprints like cuts or abrasions. The emerging technologies allow adaptability to these limitations and learning from new data, ultimately improving the systems continuously. AI-driven biometrics are extensively used in digital financial transactions by predicting fraudulent activity by analyzing behavioral patterns, namely, typing speed or navigation behaviour on a banking app.

Traditional biometrics systems lay the foundation for modern biometric systems and the integration of technologies like AI and ML have made these systems more accurate, reliable, and applicable in the real-world contexts, namely, securing financial transactions and improvising public safety. Table 1 provides the comparison of traditional and AI-based biometric systems.

Table 1: The comparison of traditional and AI-based biometric systems

Parameter/ Feature	Traditional Biometric Systems	AI-based Biometric Systems
Pattern Recognition	Rely on predefined, manual, and simple patterns	Utilize deep learning models (e.g., CNNs, RNNs) that learn complex, dynamic patterns automatically
Adaptability	Limited adaptability to changes in biometric traits (e.g., aging, injuries)	High adaptability: models learn and adjust based on new data over time
Accuracy	Generally lower accuracy, especially under varying conditions such as lighting or pose	Higher accuracy due to robust model training on vast datasets; adaptable to conditions
Real-Time Processing	Limited capability for real-time processing due to computational constraints	Enhanced real-time processing, especially for video-based applications (e.g., facial recognition)
Data Requirements	Limited data requirements, may use small datasets	Requires large datasets for training, enabling better generalization and performance
Error Rates	Higher error rates in the presence of noise or environmental variations	Lower error rates, with advanced noise reduction and filtering techniques
Learning Ability	Static; does not improve with usage or data exposure	Dynamic; improves continuously with more data and interactions
Application Areas	Primarily used in controlled environments (e.g., access control, criminal investigations)	Broad applications including mobile devices, healthcare, financial transactions, and surveillance
Response to Fraud Detection	Limited capability for detecting behavioral anomalies or advanced fraud	High capability in detecting anomalies and potential fraud by analyzing behavioral patterns

Source: Own elaboration.

3. Fundamentals of AI and Machine Learning in Biometrics

Emerging technologies including AI and ML have become an integral part of the development of biometric systems, namely, fingerprint detection, face recognition, iris scanning, and analyzing behavioral biometrics.

AI is a field of computer science that simulates human intelligence in machines. AI enables machines to perform decision-making and problem-solving tasks (Russel et al., 2016). Two types of AIare: narrow AI, which allows systems to perform a specific task like facial recognition, and general AI having capability of performing intellectual tasks like humans do. Narrow AI is employed in current biometric systems for pattern recognition.

ML is a branch of AI that is capable of building systems which can automatically learn from data without being explicitly programmed. These models are exposed to humongous amounts of data so that their performance for a specific task is enhanced over time. ML models make predictions based on patterns in the labeled or unlabeled data. They can identify unique features accurately by analyzing large sets of biometric data (e.g., fingerprints, iris scans, voice recordings) (Goodfellow, 2016).

DL is a subset of ML that utilizes artificial deep neural networks with multiple layers to model complex patterns in data. The capability of deep neural networks in learning high-level abstractions from raw data makes it very successful in applications like image processing, speech recognition, and biometric systems.

3.1. Fundamentals of Artificial Intelligence in Biometrics

Algorithms

The sophisticated algorithms in AI biometric systems analyze and interpret the biometric data. Algorithms are basically stepwise instructions for solving a problem (Sheikh et al., 2023), which in the context of biometrics is to extract features and compare fingerprints, facial structures, etc. AI algorithms enhance the accuracy of the biometric systems they continuously learn from data.

Data Processing

ML models in biometrics depend on vast amounts of data, like fingerprint images or voice recordings are pre-processed to remove noise, enhance quality, and normalize inputs. The effectiveness of any AI system in biometrics is contingent upon the quality and quantity of the training data.

Model Training

The whole purpose of ML is to train models based on historical data. In biometrics, models are trained to recognize patterns in biometric features by exposure to numerous samples. The training process allows the model to generalize from the data it has seen, enabling it to make accurate predictions with new data. In facial recognition, the system is trained on thousands of labeled face images to learn the distinguishing features of different individuals (Taigman et al., 2014).

AI and ML, in the context of biometrics, form the building blocks for systems that are more robust, adaptive, and scalable than the conventional algorithms. DL models perform amazingly well on tasks of face recognition by learning hierarchically, from edges to full faces, without manual intervention.

3.2. Core Machine Learning Techniques in Biometrics

ML is transforming biometric systems by allowing them to automatically learn patterns from biometric data and improve over time. The different types of ML models are:

Supervised Learning

The most widely used ML technique is Supervised learning. The model is trained on a labeled dataset. The training data has identity labels for each biometric record (fingerprint or face). It is trained to map the input biometric data to the corresponding label (identity) and once trained, it can make accurate predictions for unseen samples.

A supervised learning model for fingerprint recognition is trained on a dataset of labeled fingerprints in which images of fingerprints are labeled with individual identities. The model learns to identify unique patterns such as ridges and minutiae points and map fingerprint features to identities. The trained model can classify new fingerprint images based on learned patterns.

Unsupervised Learning

The unsupervised ML models are for unlabeled data. The application of unsupervised learning is clustering, anomaly detection, and organizing large biometric datasets. Face clustering is an application of unsupervised learning algorithm in which the unlabeled facial images are grouped on the basis of similarities in the features of face-like shape, texture, and distance between facial landmarks, without needing to know the identities of the individuals. This technique is highly useful in law enforcement for classifying a large collection of facial images into groups of similar-looking individuals.

Reinforcement Learning

Reinforcement learning (RL) is an advanced ML model where learning happens through interaction with an environment and the reception of feedback in the form of rewards or penalties. The decision-making process is improvised by maximizing cumulative rewards over time. The applications of RL in biometrics are optimization of processes like multi-modal biometric fusion or real-time decision-making in access control systems. RL can be used to improve the accuracy and efficiency of biometric verification in an access control system by dynamically adjusting thresholds on the basis of feedback received from successful or unsuccessful recognition attempts.

Neural Networks

Deep neural networks (DNNs) are the underlying architecture for modern biometric systems. DNNs are composed of multiple layers of neurons (nodes) that process data in a hierarchical manner. Each successive layer of the network extracts increasingly complex features from the input, making DNNs highly suitable for tasks like facial recognition, fingerprint classification, and voice verification. Figure 1 shows the neural network that takes facial images as an input for facial recognition and the first layer of the neural network detects simple features like edges or textures. The abstract features like eyes, nose, and mouth shapes are processed by the subsequent layers. The final layer of the network provides the classified output. The DL models-based facial recognition systems are more accurate and robust (Yuan et al., 2017).

Support Vector Machines (SVMs)

SVMs is another popular ML model applied in biometrics, namely, fingerprint verification, voice recognition, or iris identification. SVMs are used to find the optimal hyperplane (boundary) that separates different classes of data points and is particularly effective in classification tasks.

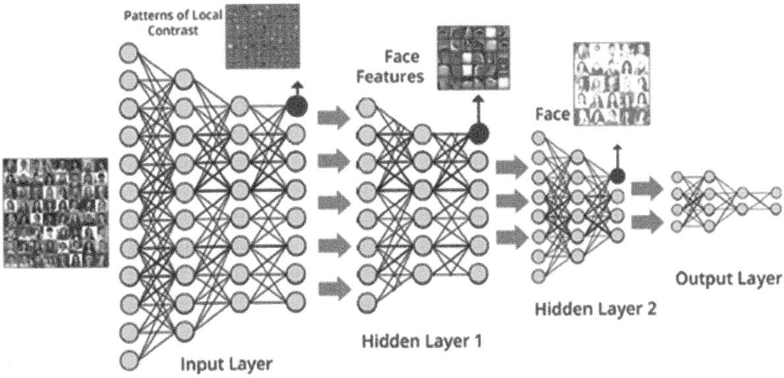

Figure 1: Neural network architecture for facial recognition.
Source: Grigsby, 2018

4. Transformative Impact of AI and ML on Biometric Modalities

4.1. Facial Recognition

Advancements in AI and ML have revolutionized facial recognition. CNNs, a class of DL algorithms, are specifically designed to automatically and adaptively learn spatial hierarchies of features from images. This capability has significantly enhanced accuracy and efficiency in facial recognition.

Traditional facial recognition systems were constrained in their capacity to handle variations in facial expressions, lighting, environment, and occlusions as they were dependent on manually designed features and statistical techniques (Roychowdhury et al., 2015). CNNs develop highly robust facial recognition systems by demonstrating exceptional proficiency in learning complex patterns from large datasets (Trigueros et al., 2018). The DL models reduce manual feature engineering as they automatically extract features from raw pixel data and improvise the system's ability to generalize across different facial images.

The practical applications of this technological advancement are prominently evident in several real-world applications. CNNs-based facial recognition systems provide a secure and user-friendly alternative to traditional password-based methods in mobile authentication. Facial recognition is an important feature of modern smartphones, tablets, etc., which are used to unlock devices, authorize transactions, and even log into applications. This integration guarantees that only authorized users can access sensitive information, and thus amplifies security.

The application of CNN-based facial recognition systems is not just limited to mobile applications. They are implemented in the field of surveillance for the real-time identification of individuals from video feeds. Facial recognition has its extensive application in airports for the verification of passengers/travelers. The use case of this is the U.S. pre-clearance and biometric entry-exit program at Dublin airport (Khan et al., 2021). U.S. Customs and Border Control (CBP) has implemented a facial recognition verification process at Dublin airport to offer pre-clearance facilities by U.S. immigration officials, leading to a significant reduction in waiting queues assuring higher customer satisfaction by faster processing systems. Most importantly, the overall security is enhanced by confirming identities at immigration and boarding gates.

4.2. Voice and Speech Recognition

RNNs, in particular, Long Short-Term Memory (LSTM) networks paved the path for developments in the voice and speech recognition detection systems. The accuracy of these systems has been enhanced due to RNNs' capability of processing and analyzing the information across long sequences (Graves et al., 2013).

DL models can capture temporal dependencies in speech and are capable of understanding context, intonation, and varying accents in speech. Voice assistants such as Apple's Siri, Amazon's Alexa, and Google Assistant are examples of advanced voice recognition systems (Hoy, 2018).

Call centers deploy voice recognition systems direct calls according to given instructions and respond automatically to commonly asked queries of customers (Devi et al., 2019). In addition to responding to frequently asked queries, these systems analyze sentiments and emotions in real-time and improve customer service by addressing the customer's needs efficiently.

They enhance security by using voiceprints for the verification of identities, thereby adding an additional layer of authentication. The most relevant implementation of this is in the financial sector, especially banks. HSBC bank allows customer authentication with the voice (Vittori, 2019). A customer is required to repeat a given phrase which is mapped against a pre-recorded voiceprint of the user and analyzed with an accuracy for ensuring the correct voice match.

4.3. Fingerprint and Palm Print Recognition

AI has greatly improved the accuracy and reliability of fingerprint and palm print recognition systems. Contrary to simple pattern-matching algorithms in traditional systems, AI models such as CNNs perform highly sophisticated analysis of fingerprint and palm print patterns. CNNs have better accuracy in matching fingerprints even in the cases of partial obscuration or degradation. This became possible by analyzing the intricate details of fingerprint ridges and minutiae points. The increased reliability of fingerprint recognition has various practical applications ranging from unlocking smartphones, laptops, and other devices to accessing secure facilities. Furthermore, the exposure to more and large datasets has improved the robustness and accuracy of palm-print recognition systems (Hemalatha, 2020).

In law enforcement, AI-enhanced fingerprint and palm print recognition systems play a vital role in criminal investigations and identification. They are used to compare prints collected from crime cities with those stored in databases, and assist in the identification and apprehension of suspects. AI-driven fingerprint and palm print recognition in secure access systems guarantee that only authorized individuals are able to enter sensitive areas, thus improving overall security (Win et al., 2020).

A unique identification system UIDAI-Aadhaar, one of the world's largest biometric databases, covering the data of more than 1.3 billion individuals, is deployed by the Government of India. It utilizes the integration of AI with fingerprint, face, and iris scanning to identify the individual and minimize fraud (Sharma et al., 2023).

The fingerprint recognition has also been in use for a long time at Disney World and Universal Studios for re-entry to parks.

4.4. Behavioral Biometrics

Behavioral biometrics enable the identification of behavioral traits, namely, voice, gait, and keystroke dynamics. The way a person walks, and non-frontal poses are the gait behavioral traits. The precise identification of individuals based on their walking patterns has become feasible because of AI-driven biometric systems. This system has its employability in tracking individuals in diverse settings for the purpose of security and surveillance. Gait analysis is used for ensuring high security in government buildings. Chinese authorities analyze body shape and way of walking to identify individuals.

The analysis of keystroke dynamics aims at identifying typing patterns like speed and rhythm for detecting fraudulent attempts. AI-driven systems can be deployed in secure login systems and fraud detection to enhance security by authenticating users based on their unique typing patterns (Alsultan et al., 2017). Barclay plc Bank of UK and CIMB bank of Malaysia and South East Asia use behavioral metrics including typing/mouse patterns to identify legitimate users for preventing fraudulent transactions, adding a layer of security for user authentication (Sijan et al., 2022; Sharma et al., 2016).

AI continuously learns from new data and fine-tunes the models for recognizing individual behavior with a higher accuracy, refines, and enhances these behavioral biometrics. Consequently, this leads to decreased false acceptance/rejection rates and improved accuracy making behavioral biometrics a very promising field for future developments in security and user authentication.

To summarize, the accuracy, efficiency, and applicability of biometric modalities has been highly improvised by the integration of AI and ML. Facial, voice and speech, fingerprint and palm print recognition systems, and behavioral biometrics systems have become more secure and user-friendly due to advancements in these technologies. Security enhancement and personal authentication will see developments with the ongoing evolution and innovation in AI and biometrics.

5. Applications Across Key Sectors

Biometric systems have been extensively used in several prominent sectors.

5.1. Banking and Financial Services

AI-driven biometrics have enhanced security and simplified the digital transaction processes in the banking and financial sector. The integration of biometric systems is imperative for protecting digital transactions and combating fraud (Singh et al., 2019). In contrast to biometric authentication, traditional methods of authentication, namely, passwords, tokens, and PINs can be forgotten or stolen, and hence are less secure. Facial recognition and fingerprint scanning are being deployed in mobile banking apps for secured account access and transaction approval without remembering complex passwords. Thus, biometrics lead to security and improved user-convenience and engagement.

By analyzing biometric data in association with transaction patterns, banks can detect unusual activities and prevent fraudulent transactions more effectively (Soares et al., 2016). JP Morgan Chase & Co Bank, China deploys multiple user authentication methods, like fingerprint or facial recognition and voiceprint authentication. This allows the users a secure access to their accounts (Jagadeesh et al., 2023).

The implementation of biometric systems in banking extends to ATM security. Several banks have decreased the risk of card skimming and unauthorized access by introducing biometric ATMs equipped with fingerprint sensors and facial recognition cameras (Agidi, 2018). Biometrics systems have a crucial role in the e-commerce and e-banking sectors. The access of e-commerce applications is growing rapidly leading to a surge in online transactions including payments and e-banking operations. Despite being a small fraction of total transactions, the online transactions account for a significant share of losses occurred by financial institutions (Scaria et al., 2018; Khan, 2014). AI-based Biometric systems are increasingly applied to enhance the security of one-time passwords for e-commerce and e-banking transactions.

5.2. Healthcare

There is a high impact of biometric systems in the healthcare domain. AI integrated facial recognition and other biometrics modalities are used for patient identification and authentication. A patient's identity is accurately recognized by biometrics, thus reducing the risk of any errors or confusions in providing the right treatment to the right individual. Additionally, AI-based biometrics can be applied for detecting diseases.

Since medical information is quite sensitive, AI-driven biometrics must be deployed to protect patient confidentiality. It is feasible by creating highly secure and personalized access control systems. Voice recognition technology is applied for accessing electronic health record (EHR) systems, allowing for secure and hands-free access to patient information (Andriole, 2014).

Binary classification task of classifying healthy and diseased persons on the basis of biometric data is performed using SVMs (Chen et al., 2023; Lynch et al., 2018).

5.3. Law Enforcement and Public Security

AI-driven biometric systems have revolutionized law enforcement and public security by providing enhanced capabilities in surveillance, forensics, and criminal identification. AI-powered facial recognition systems enable real-time monitoring of public spaces, allowing law enforcement agencies to quickly identify individuals of interest and detect suspicious activities in the context of surveillance (Smith et al., 2021). The accurate and speedy analysis of fingerprints and DNA matching in forensic applications has proven to be instrumental in criminal identification and collection of evidence for solving high profile cases (Bond, 2009).

6. Challenges and Ethical Considerations

The ethical and privacy concerns are being raised by the application of AI-enhanced biometric systems. With the advancement in these technologies, it is very important to address the associated ethical and privacy concerns to ensure their responsible and equitable implementation (Dhirani et al., 2023). Policies and regulations are needed for the responsible use of AI technologies in biometric systems to ensure data security.

6.1. Bias and Fairness in AI-Driven Biometrics

AI-driven biometrics raise the issues of bias and fairness. The training datasets are reasons for these issues in the models. The model will not classify the underrepresented groups accurately if there is a lack of representation of all demographic groups in the training data, leading to discrimination in results (Cherepanova et al., 2022). AI models can reflect societal biases present in the biometric data (Ferrara, 2023). The concerns of equitability and inclusiveness are evident in identifying individuals with darker tones or specific gender traits inaccurately (Mujtaba et al., 2019). These biases have serious technical and societal consequences. Biased facial recognition systems can impact fairness in law enforcement and privacy by identifying the individuals incorrectly.

6.2. Approaches to Mitigate Bias and Ensure Fairness

The ethical challenges of fairness and bias can be mitigated through a multifaceted approach. The datasets which represent the demographic and socioeconomic diversity of the target population must be used for training the models (Drozdowski et al., 2020). The fairness-aware algorithms have been developed for the fair and robust biometric systems (Chouldechova et al., 2018). AI-driven biometric systems must be designed and deployed considering underrepresented groups. The stakeholder engagement and accounting of the concerns of different communities will facilitate in creating ethical and equitable biometric systems.

6.3. Privacy and Data Protection

The challenges of collecting and storing biometric data are data protection and privacy. Biometric data includes fingerprints, facial images, etc., which are private characteristics unique to a person, hence at a high risk of misuse if not controlled. The conceivable misuses of biometric data are unauthorized access and data breaches which have biometric data can result in very harmful effects because, unlike passwords, biometric traits cannot change (Rane et al., 2013; Mordini, 2009).

6.4. Data Management Strategies

These strategies are required to tackle all these issues and for secured data management. Security can also be achieved by offering strong encryption methods on the biometric data during its transmission and storage process. Encryption ensures that it is impossible to read it, even though intercepted or accessed unauthorizedly, unless an authorized key for decryption is applied (Gupta et al., 2018). Biometric information needs further securing using secure servers and databases with tight access controls. The organizations ought to use these standard security audits and vulnerability assessments in determining and mitigating possible weaknesses in security.

The other important aspect of data protection is explicit user consent. Notwithstanding, the users should know when their biometric data is acquired, used, and stored, and thus have given their consent before its collection. (Jayapal et al., 2016). Such transparency in utilization should be ensured, and control over the individual's data should always be held by the individual for privacy and trust to be upheld.

6.5. Regulatory Guidelines

Under the General Data Protection Regulation of the European Union, there exists an all-rounded framework for the protection of biometric data. The GDPR requires that there be consent from the individuals before one gathers the biometric data and that the data be processed according to law, fairly, and transparently (European Commission, 2016) (Sanchez-Reillo et al., 2019). Another regulation to consider is the California Consumer Privacy Act passed in the United States. This regulation protects the use of biometric data, in a similar manner as GDPR, whereby it focuses on security in relation to data, user consent, and even privacy rights on access and deleting information (California Legislative Information, 2018) (Determann et al., 2018).

This worldwide regulatory landscape for biometric technologies, therefore, is pretty diversified and different from one jurisdiction to another. It seems that this is a major challenge in businesses implementing biometric systems since proper comprehension of regulation compliance is pretty daunting.

There are many countries that have enacted laws regulating the use of biometric technologies. For instance, the EU's GDPR set out definite rules on processing biometric data with data protection impact assessments and data subject rights among others (European Commission, 2016). On its part, the Personal Information Protection Law of China outlined clear regulations regarding the collection and processing of personal data, like biometric data, where the individual's privacy will be protected (China Law Translate, 2021) (Cui et al., 2021).

In the United States, regulation is very fragmented, and biometric data protection has been covered through state legislation such as the Illinois Biometric Information Privacy Act (BIPA), which imposes specific requirements on the collection and storage of biometric data (González-Pizarro et al., 2022). This makes compliance challenging. There are also several burdens in complying with these requirements for businesses. For instance, businesses face the challenge of dealing with the various and diversified needs of the jurisdictions. The international firms must ensure that the frameworks of biometric data practices are aligned to the regulatory structure of each country, an exercise which can be very complex and very resource intensive. In addition, changing rules and regulations continually demand real-time monitoring and updating of policies related to data protection. As the rules and regulations concerning biometric data are constantly evolving, it leaves businesses in dire need of being updated with new legislation requirements and their implementation in the policies and practices.

6.6. Legal Frameworks and Commentaries

Legal frameworks such as GDPR and BIPA set forth the minimum standard that must be followed in the protection of biometric data. Commentaries by authors such as International Data Privacy Law describe in detailed terms what the law says, its impact in terms of compliance among businesses and individuals, challenges of compliance with regulatory change, and best practices on managing biometric data to meet legal requirements (Bustard, 2015).

In conclusion, there is a great need to take up the challenges and ethical considerations that AI-driven biometrics lay out so that the technologies are used responsibly and equitably. This can be done by mitigating bias, safeguarding privacy, and meeting regulatory expectations so that there is an assurance of the harnessing of biometric technology to an individual's advantage without necessarily being opposed to individual rights and without being unfair (Bacchini et al., 2019).

7. Future Trends in AI and ML for Biometrics

This has led to the fast growth of Biometric technologies regarding the increasing risks to information security as they have proved to be a reliable and secure form of identification and verification (Alharbi et al., 2023). Biometrics has also been increasingly entwined with the advancements in AI and ML in the era of digitalization. The new technologies have revamped how biometric systems work and open the door to better secure accurate, and efficient systems (Wayman, 2007).

7.1. Frictionless Biometrics

One of the emerging domains for the future is "frictionless biometrics", a specialty area of behavioral biometrics (Genovese et al., 2018). Dynamic behavioral attributes like typing patterns, gait, voice modulations, and manners are significant shifts from the physical biometric attributes like fingerprints, etc., to be used for identification. These dynamic attributes cannot be mimicked and can be effortlessly drawn from individuals without requiring them to consciously participate in the process and therefore promise the enhancement of security and user experience. The advancement of 'frictionless' authentication opens up possibilities for 'invisible' authentication, streamlining access control and enhancing user experience.

7.2. Integration with Emerging Technologies

Biometric systems, particularly when integrated into emerging technologies like blockchain and the Internet of Things (IoT), present one of the bright vistas for producing secure, future-proof biometric systems. Strong infrastructure for the security of biometric data is provided by blockchain technology that is famed for its decentralized and tamper-proof ledger. Immunizing and protecting against undesirable modification can be ensured by the businesses through encrypting biometric data on a blockchain, which can enhance the integrity of biometric systems by encouraging secure and clear data sharing across different entities (Pawade et al., 2019).

Biometrics also enhance security and convenience. IoT devices, smart locks, or wearable technology can make use of biometric authentication to provide safe access or personalized experience.

Moreover, IoT-enabled biometric systems are capable of retrieving and processing data from various sensors, leading to a comprehensive authentication of an individual as it accounts for environmental and contextual factors into the data. The application of these integrations is in the domain of home automation and health monitoring (Wójtowicz et al., 2016).

8. Conclusion

The integration of AI and ML has radically made the advancements in the field of biometrics by enhancing security and user experience across applications. This chapter presents different aspects of AI and ML in developing biometric systems by discussing leading trends, challenges, and future developments in the field. The rapid advancements and innovations in technology are continuously evolving biometric systems, and hence this domain needs to be researched for improving the identity verification and authentication processes.

The combination of several biometric modalities with behavioral traits is shifting the landscape of user authentication to an advanced and user-friendly system. The employability of AI in making personalized biometric systems based on the user's preferences is in trend. This trend in personalized and adaptive authentication methods is meant to help in providing better security and satisfaction through more appropriate and accurate authentications (Zhu et al., 2022).

In the near future, expected progress in AI and biometrics encompass the development of multimodal biometric systems where varied biometric modalities, namely, facial recognition, fingerprint scanning, and voice recognition, are integrated into one authenticating procedure (Naidu et al., 2019). It is bound to be more accurate and secure because it exploits the benefits offered by several techniques of (Bah et al., 2021). Further research is expected to lead to advancements in real-time biometric identification systems, which are more functional in various and dynamically changing environments. Advances in AI algorithms-federated learning and edge computing, for instance, would play a crucial role in realizing such breakthroughs as they enhance both the efficiency and the scalability of the systems with which biometrics is used.

References

Agidi, R.C. (2018). Biometrics: The future of banking and financial service industry in Nigeria. *International Journal of Electronics and Information Engineering*, 9(2), 91–105.

Albalawi, S., Alshahrani, L., Albalawi, N., Kilabi, R., & Alhakamy, A.A. (2022). A comprehensive overview on biometric authentication systems using artificial intelligence techniques. *International Journal of Advanced Computer Science and Applications*, 13(4), 1–11.

Alharbi, B., & Alshanbari, H.S. (2023). Face-voice based multimodal biometric authentication system via FaceNet and GMM. *PeerJ Computer Science*, 9, e1468.

Alsultan, A., Warwick, K., & Wei, H. (2017). Non-conventional keystroke dynamics for user authentication. *Pattern Recognition Letters*, 89, 53–59.

Alzamel, H.A., Al-Shabanah, M., & Alsmadi, M. (2019). Point of sale (POS) network with embedded fingerprint biometric authentication. *International Journal of Scientific Research in Science and Technology* (*IJSRST*), Online ISSN.

Andriole, K.P. (2014). Security of electronic medical information and patient privacy: What you need to know. *Journal of the American College of Radiology*, 11(12), 1212–1216.

Bacchini, F., & Lorusso, L. (2019). Race, again: How face recognition technology reinforces racial discrimination. *Journal of Information, Communication, and Ethics in Society*, 17(3), 321–335.

Bah, C.U., Seyal, A.H., & Yahya, U. (2021). Combining pin and biometric identifications as enhancement to user authentication in internet banking. arXiv preprint arXiv:2105.09496.

Bhatt, H.S., Bharadwaj, S., Singh, R., & Vatsa, M. (2010, January). Face recognition and plastic surgery: Social, ethical, and engineering challenges.

In: *International Conference on Ethics and Policy of Biometrics* (pp. 70–75). Berlin, Heidelberg: Springer Berlin Heidelberg.

Bond, J.W. (2009). The value of fingerprint evidence in detecting crime. *International Journal of Police Science & Management*, 11(1), 77–84.

Bustard, J. (2015). The impact of EU privacy legislation on biometric system deployment: Protecting citizens but constraining applications. *IEEE Signal Processing Magazine*, 32(5), 101–108.

Campbell, J.P. (1997). Speaker recognition: A tutorial. In: *Proceedings of the IEEE*, 85(9), 1437–1462.

Chen, F., Lu, H.L., & Simonetti, A. (2023). Practical statistical considerations for the clinical validation of AI/ML-enabled medical diagnostic devices. arXiv preprint arXiv:2303.05399.

Cherepanova, V., Reich, S., Dooley, S., Souri, H., Dickerson, J., Goldblum, M., & Goldstein, T. (2023, August). A deep dive into dataset imbalance and bias in face identification. In: *Proceedings of the 2023 AAAI/ACM Conference on AI, Ethics, and Society* (pp. 229–247).

Chouldechova, A., & Roth, A. (2018). The frontiers of fairness in machine learning. arXiv preprint arXiv:1810.08810.

Cui, S., & Qi, P. (2021). The legal construction of personal information protection and privacy under the Chinese Civil Code. *Computer Law & Security Review*, 41, 105560.

Daugman, J. (2001, October). High confidence recognition of persons by iris patterns. In: *Proceedings IEEE 35th Annual 2001 International Carnahan Conference on Security Technology* (Cat. No. 01CH37186) (pp. 254–263). IEEE.

Daugman, J. (2009). How iris recognition works. In: *The Essential Guide to Image Processing* (pp. 715–739). Academic Press.

Determann, L. (2018). New California Law Against Data Sharing: The California Consumer Privacy Act of 2018, Broad Data and Business Regulation, Applicable Worldwide. *Computer Law Review International*, 19(4), 117–124.

Devi, S., Merchant, Z.A.M., Siddiqui, M.S., & Lobo, M. (2019). Artificial intelligence based personal assistant. *Asian Journal for Convergence in Technology (AJCT)*, ISSN-2350-1146.

Dey, A., Tharmavaram, M., Pandey, G., Rawtani, D., & Mustansar Hussain, C. (2020). Conventional and emerging biometrics techniques in forensic investigations. *Technology in Forensic Science: Sampling, Analysis, Data, and Regulations*, 175–197.

Dhirani, L.L., Mukhtiar, N., Chowdhry, B.S., & Newe, T. (2023). Ethical dilemmas and privacy issues in emerging technologies: A review. *Sensors*, 23(3), 1151.

Drozdowski, P., Rathgeb, C., Dantcheva, A., Damer, N., & Busch, C. (2020). Demographic bias in biometrics: A survey on an emerging challenge. *IEEE Transactions on Technology and Society*, 1(2), 89–103.

Ferrara, E. (2023). Fairness and bias in artificial intelligence: A brief survey of sources, impacts, and mitigation strategies. *Sci*, 6(1), 3.

Genovese, A., Munoz, E., Piuri, V., & Scotti, F. (2018). Advanced biometric technologies: Emerging scenarios and research trends. *From Database to Cyber Security: Essays Dedicated to Sushil Jajodia on the Occasion of His 70th Birthday*, 324–352.

González-Pizarro, F., Figueroa, A., López, C., & Aragon, C. (2022). Regional differences in information privacy concerns after the Facebook-Cambridge Analytica data scandal. *Computer Supported Cooperative Work (CSCW)*, 31(1), 33–77.

Goode, A. (2018). Biometrics for banking: Best practices and barriers to adoption. *Biometric Technology Today*, 2018(10), 5–7.

Goodfellow, I., Bengio, Y., & Courville, A. (2016). *Deep Learning*. Adaptive Computation and Machine Learning Series. MIT Press.

Graves, A., Mohamed, A.R., & Hinton, G. (2013, May). Speech recognition with deep recurrent neural networks. In: *2013 IEEE International Conference on Acoustics, Speech and Signal Processing* (pp. 6645–6649). IEEE.

Grigsby, S.S. (2018). Artificial intelligence for advanced human-machine symbiosis. In: *Augmented Cognition: Intelligent Technologies: 12th International Conference, AC 2018*, Held as Part of HCI International 2018, Las Vegas, NV, USA, 15–20 July. *Proceedings, Part I* (pp. 255–266). Springer International Publishing.

Gupta, H., & Anne-Marie, C.A.A. (2018). Role of multiple encryptions in biometric devices. In: *Cyber Security: Proceedings of CSI 2015* (pp. 291–300). Springer Singapore.

Hemalatha, S. (2020, February). A systematic review on fingerprint based biometric authentication system. In: *2020 International Conference on Emerging Trends in Information Technology and Engineering (IC-ETITE)* (pp. 1–4). IEEE.

Hoy, M.B. (2018). Alexa, Siri, Cortana, and more: An introduction to voice assistants. *Medical Reference Services Quarterly*, 37(1), 81–88.

Jagadeesh, S., Ali, S.M., Selvan, S.P.G., Aljanabi, M., & Gopianand, M. (2023). Hybrid AES-modified ECC algorithm for improved data security over Cloud storage. *Journal of Advanced Research in Applied Sciences and Engineering Technology*, 32(1), 46–56.

Jain, A.K. (2005). Biometric recognition: How do I know who you are? In: *Image Analysis and Processing–ICIAP 2005: 13th International Conference,* Cagliari, Italy, 6–8, September *Proceedings 13* (pp. 19–26). Springer Berlin Heidelberg.

Jain, A.K., Nandakumar, K., & Ross, A. (2016). 50 years of biometric research: Accomplishments, challenges, and opportunities. *Pattern Recognition Letters*, 79, 80–105.

Jayapal, R., & Govindan, P. (2016, October). Biometric encryption system for increased security. In: *2016 IEEE International Carnahan Conference on Security Technology (ICCST)* (pp. 1–3). IEEE.

Khan, H.U. (2014). Research article E-banking: Online transactions and security measures. *Research Journal of Applied Sciences, Engineering, and Technology*, 7(19), 4056–4063.

Khan, Nimra, & Marina Efthymiou. (2021). The use of biometric technology at airports: The case of customs and border protection (CBP). *International Journal of Information Management Data Insights*, 1(2), 100049. http://dx.doi.org/10.1016/j.jjimei.2021.100049

Kortli, Y., Jridi, M., Al Falou, A., & Atri, M. (2020). Face recognition systems: A survey. *Sensors*, 20(2), 342.

Lynch, C.J., & Liston, C. (2018). New machine-learning technologies for computer-aided diagnosis. *Nature Medicine*, 24(9), 1304–1305.

Mahmoud, A.T., Awad, W.A., Behery, G., Abouhawwash, M., Masud, M., Aljuaid, H., & Ebada, A.I. (2023). An automatic deep neural network model for fingerprint classification. *Intell. Autom. Soft Comput.*, 36, 2007–2023.

Modi, K., & Devaraj, L. (2022). Advancements in biometric technology with artificial intelligence. arXiv preprint arXiv:2212.13187.

Mordini, E. (2009). Ethics and policy of biometrics. In: *Handbook of Remote Biometrics: For Surveillance and Security* (pp. 293–309). London: Springer.

Muhammad, J., Wang, Y., Wang, C., Zhang, K., & Sun, Z. (2021). Casia-face-africa: A large-scale African face image database. *IEEE Transactions on Information Forensics and Security*, 16, 3634–3646.

Mujtaba, D.F., & Mahapatra, N.R. (2019, November). Ethical considerations in AI-based recruitment. In: *2019 IEEE International Symposium on Technology and Society (ISTAS)* (pp. 1–7). IEEE.

Naidu, B.R., Bhavani, K.V.L., Rao, C.S., & Reddy, P.P. (2019, April). Comparative analysis of three single trait biometric authentication models. In: *2019 International Conference on Communication and Signal Processing (ICCSP)* (pp. 0956–0959). IEEE.

Pampus, J., & Weber, F. (1998). Facial recognition—An overview. *Information Security Technical Report*, 3(1), 40–46.

Pawade, D., Sakhapara, A., Andrade, M., Badgujar, A., & Adepu, D. (2019). Implementation of fingerprint-based authentication system using blockchain. In: *Soft Computing and Signal Processing: Proceedings of ICSCSP 2018*, Volume 1 (pp. 233–242). Springer Singapore.

Rane, S., Wang, Y., Draper, S.C., & Ishwar, P. (2013). Secure biometrics: Concepts, authentication architectures, and challenges. *IEEE Signal Processing Magazine*, 30(5), 51–64.

Roychowdhury, S., & Emmons, M. (2015). A survey of the trends in facial and expression recognition databases and methods. arXiv preprint arXiv:1511.02407.

Russell, S.J., & Norvig, P. (2016). *Artificial Intelligence: A Modern Approach.* Pearson. London.

Sanchez-Reillo, R., Ortega-Fernandez, I., Ponce-Hernandez, W., & Quiros-Sandoval, H.C. (2019). How to implement EU data protection regulation for R&D in biometrics. *Computer Standards & Interfaces*, 61, 89–96.

Scaria, B.A., & Megalingam, R.K. (2018, June). Enhanced E-commerce application security using three-factor authentication. In: *2018 Second International Conference on Intelligent Computing and Control Systems (ICICCS)* (pp. 1588–1591). IEEE.

Sharma, A., & Lenka, S.K. (2016). E91 QKD protocol for authentication in online banking systems. *International Journal of Business Information Systems*, 22(1), 116–122.

Sharma, D., & Selwal, A. (2023). A survey on face presentation attack detection mechanisms: Hitherto and future perspectives. *Multimedia Systems*, 29(3), 1527–1577.

Sheikh, H., Prins, C., & Schrijvers, E. (2023). Artificial intelligence: Definition and background. In: *Mission AI: The New System Technology* (pp. 15–41). Cham: Springer International Publishing.

Sijan, M.A.H., Shahoriar, A., Salimullah, M., Islam, A.S., & Khan, R.H. (2022, March). A review on e-banking security in Bangladesh: An empirical study. In: *Proceedings of the 2nd International Conference on Computing Advancements* (pp. 330–336). Dhaka, Bangladesh.

Singh, A., Srivastva, R., & Singh, Y.N. (2019). Prevention of payment card frauds using biometrics. *International Journal of Recent Technology and Engineering (IJRTE)*, 8, 516–525.

Smith, M., & Miller, S. (2022). The ethical application of biometric facial recognition technology. *AI & Society*, 37(1), 167–175.

Soares, J., & Gaikwad, A.N. (2016, September). Fingerprint and iris biometric controlled smart banking machine embedded with GSM technology for OTP. In: *2016 International Conference on Automatic Control and Dynamic Optimization Techniques (ICACDOT)* (pp. 409–414). IEEE.

Sreeja, S.P., Sahu, M.K., & Hiremanth, N. (2018). A survey on biometrics security systems. *Int. J. Recent Trends Eng. Res.*, 4(3). 263-268.

Taigman, Y., Yang, M., Ranzato, M.A., & Wolf, L. (2014). Deepface: Closing the gap to human-level performance in face verification. In: *Proceedings of the IEEE Conference on Computer Vision and Pattern Recognition* (pp. 1701–1708). IEEE.

Trigueros, D.S., Meng, L., & Hartnett, M. (2018). Face recognition: From traditional to deep learning methods. arXiv preprint arXiv:1811.00116.

Vittori, P. (2019). Ultimate password: Is voice the best biometric to beat hackers? *Biometric Technology Today*, 2019(9), 8–10.

Wayman, J.L. (2007). The scientific development of biometrics over the last 40 years. In: *The History of Information Security* (pp. 263–274). Elsevier Science BV.

Win, K.N., Li, K., Chen, J., Viger, P.F., & Li, K. (2020). Fingerprint classification and identification algorithms for criminal investigation: A survey. *Future Generation Computer Systems*, 110, 758–771.

Wójtowicz, A., & Joachimiak, K. (2016). Model for adaptable context-based biometric authentication for mobile devices. *Personal and Ubiquitous Computing*, 20(2), 195–207.

Wu, Y., & Ji, Q. (2019). Facial landmark detection: A literature survey. *International Journal of Computer Vision*, 127(2), 115–142.

Yuan, L., Qu, Z., Zhao, Y., Zhang, H., & Nian, Q. (2017, March). A convolutional neural network based on TensorFlow for face recognition. In: *2017 IEEE 2nd Advanced Information Technology, Electronic, and Automation Control Conference* (*IAEAC*) (pp. 525–529). IEEE.

Zhu, G., & Al-Qaraghuli, Y. (2022). AI-assisted authentication: State of the art, taxonomy, and future roadmap. arXiv preprint arXiv:2204.12492.

2

Biometrics as a Key Factor for Emergent User Experience Evaluation

Juan Antonio Torrecilla-García*(0000-0002-4179-0008) **and Agnieszka Grazyna Skotnicka**(0000-0001-5676-6687)
University of Málaga, Spain

1. Introduction

In an era marked by digital transformation, biometric technologies have transcended their initial applications in finance to become pivotal in enhancing user experiences across various sectors. This conceptual research and case study explores the role of biometrics as a UX evaluator, with special focus on its implementation in diverse industries like retail, healthcare, and customer service.

Biometrics, the science of identifying individuals based on their physiological and behavioral characteristics, has emerged as a pivotal tool in enhancing user experience evaluation. Traditional UX assessments often rely on self-reports, human-system performance metrics, and observational data, which may not fully capture the user's psychological states and affective responses. This theoretical background explores how biometric technologies can provide richer, more accurate insights into UX, thereby addressing the limitations of conventional methods.

Traditionally, UX evaluation relied heavily on self-reported data, including surveys and interviews, which are susceptible to various biases and limitations (Huseynov et al., 2019). The advent of sophisticated biometric technologies has enabled researchers to capture objective,

Corresponding author: juantorrecilla@uma.es

physiological responses that often elude conscious awareness or verbal articulation. These technologies include eye-tracking, facial expression analysis, galvanic skin response (GSR), fingerprint recognition, and electroencephalography (EEG), among others (Cowley et al., 2016).

The incorporation of biometrics in UX evaluation offers several advantages. (1) It provides real-time, continuous data collection throughout the user's interaction with a product or interface, allowing for a granular analysis of the user journey (Mauss & Robinson, 2009). (2) Biometric data can reveal subconscious reactions and micro-expressions that users may be unaware of or able to report accurately, thus offering a more comprehensive understanding of the user's emotional and cognitive states (Ekman & Friesen, 1969).

Traditional UX evaluation methods, like self-reports and observational data, have significant limitations. Self-reports require users to articulate their subjective experiences, which can be challenging due to a lack of self-awareness or reluctance to express true feelings. Observational methods necessitate analysts to infer psychological states from external behaviors, which may not always be accurate. These methods often fail to capture the nuanced psychological demands and affective responses of users, leading to incomplete UX assessments.

The application of biometrics in UX evaluation extends across various domains, including web design, mobile applications, virtual and augmented reality experiences, and physical product interactions. In these contexts, biometric data has proven invaluable in identifying pain points, optimizing user flows, and enhancing overall user satisfaction (Hassenzahl & Tractinsky, 2006).

As we delve deeper into the era of personalized and adaptive user interfaces, the role of biometrics in UX evaluation becomes even more critical. Machine learning (ML) algorithms can leverage biometric data to create dynamic, responsive systems that adapt in real-time to users' emotional and cognitive states, potentially revolutionizing human-computer interaction (Picard, 1997).

This paper aims to provide a comprehensive overview of the current state of biometric integration in UX evaluation, exploring its methodologies, challenges, and future directions. By examining case studies and empirical research, we will demonstrate how biometrics are reshaping our understanding of UX and driving innovation in interface design and product development.

The combination of biometric data with traditional UX research methods has led to the development of more robust and holistic evaluation frameworks. This multi-method approach, often referred to as triangulation, enhances the validity and reliability of UX assessments by corroborating findings across different data sources (Jick, 1979). For instance, eye-tracking data is used to validate self-reported attention patterns, while GSR measurements provide objective indicators of stress/ excitement levels during specific interactions.

2. Theoretical Background

The integration of biometrics into UX evaluation has emerged as a critical factor in enhancing the interaction between users and technology. Biometrics offers a promising avenue for improving security and usability in various applications, particularly in mobile and online environments. This theoretical background will explore the role of biometrics in UX evaluation, highlighting its implications for usability, user satisfaction, and security. Biometric technologies offer a promising solution to the limitations of traditional UX evaluation methods. By measuring physiological and behavioral indicators, biometric systems provide objective data on users' psychological states and affective responses, which can supplement traditional methods, offering a comprehensive understanding of UX (Kumar & Zhang, 2010).

Eye tracking can reveal where users focus their attention, indicating areas of interest or confusion in a user interface. EEG can measure brain activity, providing insights into users' cognitive load and emotional states during interactions with a system (Schiavone et al., 2019; Zhang et al., 2009). These biometric indicators can help researchers and designers understand the underlying factors influencing UX, leading to user-centered design improvements. Biometric systems, like facial recognition, fingerprint scanning, and keystroke dynamics, have shown to significantly enhance UX by providing seamless and intuitive authentication methods. Kloet and Yang (2022) emphasize that multimodal biometric authentication positively influences the overall UX, even if it does not enhance specific user perceptions significantly. This aligns with Blanco-Gonzalo et al. (2014), who argue that improving user interaction with biometric systems is essential to prevent rejection and ensure optimal performance. The usability of biometric systems is paramount, as a lack of usability can lead to user frustration and decreased adoption rates, hence user acceptance must be considered.

Studies have shown that users' perceptions of biometric systems can significantly influence their overall experience, acceptance, and satisfaction. For example, a study comparing keystroke and face recognition systems found that users were more satisfied with the keystroke system, despite its lower performance metrics, suggesting that factors like ease of use, perceived performance, and robustness against attacks play a crucial role in user acceptance (El-Abed et al., 2012; Kukula et al., 2010; Theofanos et al., 2007). Demographic factors including s age and gender can influence users' perceptions of biometric technologies. Understanding these factors is essential for designing biometric systems that cater to diverse user groups and enhance overall user satisfaction (Fernandez-Saavedra et al., 2010; Yamazaki, 2019). The ethical considerations surrounding biometric data collection are crucial in the context of UX. Idoko (2024) discusses the dual challenge of utilizing biometric data for enhanced security while protecting customer privacy, highlighting the potential for invasive surveillance. This ethical dilemma underscores the necessity for designers to balance security enhancements with user trust and privacy concerns, the vital components of a positive UX.

The performance of biometric systems is directly related to the quality of the sensed data. A novel approach to improving biometric authentication performance involves quantifying the quality of user templates and incorporating it into the matching process. By extracting user quality based on the confidence of generating reliable matching scores, researchers have demonstrated significant improvements in biometric system performance (Sarkar & Singh, 2020). This approach highlights the importance of considering user-specific factors in the design and evaluation of biometric systems. The interaction between users and biometric sensors is a critical factor in the overall performance and usability of biometric systems. Traditional system-level performance metrics, including failure-to-acquire (FTA) and false-reject rates (FRR), often overlook the nuances of human-sensor interaction. The Human-Biometric-Sensor Interaction (HBSI) evaluation method addresses this gap by incorporating ergonomics, usability, and sample quality criteria into the evaluation process (Brockly et al., 2015; Elliott & Kukula, 2010). The HBSI method segments the systemic FTA rate into metrics like false interaction (FI), failure to detect (FTD), and concealed interaction (CI). This detailed analysis helps identify whether performance issues are due to the system, the user, or other factors, providing valuable insights for improving biometric system design and usability (Corsetti et al., 2020).

Biometric system performance can be influenced by various human factors, including behaviour, cognition, and anthropometric qualities.

Studies have shown that feedback, habituation, and the physical setup of biometric sensors can affect image quality and error rates in fingerprint recognition systems. Understanding these factors is essential for reducing biometric systems' uncertainty and improving their reliability. Developing robust evaluation methodologies is crucial for assessing their usability and performance. A general methodology based on the International Standard ISO/IEC 19795-2 has been proposed to analyze end-to-end system performance under different usability conditions. This methodology considers biometric characteristic variability due to illness or climatic changes and provides a controlled scenario for conducting trials. Such comprehensive evaluation frameworks are essential for producing intercomparable results and advancing the field of biometric UX evaluation.

Biometrics leverages unique physiological and behavioral traits for identification and authentication, providing a seamless and secure user interaction (Brostoff, 2017). These technologies simplify processes, reduce cognitive load, and enhance security by eliminating the need for passwords and PINs. Facial recognition and voice biometrics enable instantaneous access and personalized services, thus improving overall user satisfaction and trust (Abdulkareem & Gordon, 2023; Koonsanit & Nishiuchi, 2020). Continuous authentication systems, which utilize biometric data for verification without explicit user actions, represent an innovative approach to enhancing UX. Ryu et al. (2023) highlight that implicit authentication can significantly improve UX by operating in the background, thus minimizing disruptions. This method aligns with the findings of Schiavone et al. (2019) who advocate for multi-biometric systems that provide continuous authentication, thereby enhancing security without compromising usability. Such systems can adapt to user behavior, offering a more personalized and less intrusive experience.

Some authors (Gudala et al., 2022; Sarier, 2022; Toutara & Spathoulas, 2020) explore the integration of biometric authentication with blockchain technology to enhance the security of Identity and Access Management (IAM) systems. Conventional password-based methods have demonstrated significant vulnerabilities, including susceptibility to phishing attacks, credential stuffing, and brute-force attacks. Biometric authentication, which uses unique biological characteristics like fingerprints, etc., offers a more robust solution. However, the centralized storage of biometric data poses security risks, as a breach could expose sensitive information.

Blockchain technology presents a potential solution to these challenges by decentralizing data storage. The immutability and cryptographic underpinnings of blockchain ensure tamper-proof data management, mitigating the risks associated with centralized systems. By storing cryptographic hashes of biometric data on the blockchain, user privacy is preserved, and access control becomes more secure through the use of permissioned smart contracts. The decentralized structure of blockchain removes vulnerabilities caused by single points of failure, enhancing the system's resistance to cyberattacks. The integration of blockchain also streamlines access control processes via smart contracts, which automatically manage privileges based on predefined rules, enhancing both efficiency and security. Despite its benefits, challenges related to scalability and performance—such as blockchain's transaction speed and latency—remain significant hurdles that need to be addressed. The study concludes that the combination of biometrics and blockchain offers a promising avenue for developing more secure IAM systems, though further research is needed to optimize performance and address implementation challenges.

2.1. The Emergence of Biometric Sensors in UX Evaluation

Biometric sensors are redefining how UX is understood and measured. The research by Kula et al. (2019) and Iriarte et al. (2021) explored the potential of biometric sensors, highlighting their ability to detect subtle physiological changes that traditional UX evaluation methods often miss. These sensors include EEG, eye tracking, and facial expression analysis, each of which provides a unique window into the user's emotional and cognitive states during their interaction with products or digital interfaces.

Eye Tracking

Eye-tracking technology, one of the most established biometric methods, is widely used to monitor where users are looking, how long they focus on specific elements, and the patterns of their gaze. It helps determine the visual appeal, clarity, and navigability of interfaces or physical products. According to Salminen et al. (2009), eye tracking can be used to identify specific areas of interest (AOIs) and visual engagement, which are critical in optimizing design elements like website layouts and product packaging. Eye tracking is also often combined with other methods, such as EEG and facial analysis, to enhance the depth of data collection (Lasa et al., 2015).

Electroencephalography (EEG)

EEG measures electrical activity in the brain, offering insights into cognitive load, attention, and emotional responses. EEG has proven useful in identifying users' emotional reactions to design elements or gameplay experiences, as highlighted by Nacke's study (2009). These brainwave patterns indicate frustration, engagement, or confusion, providing designers with valuable feedback that can inform product refinement. Advances in EEG technology have allowed for more portable and user-friendly devices, making it easier to incorporate this method into real-time UX studies.

Facial Expression Analysis

Facial expression analysis (FEA) uses algorithms to detect micro-expressions and other subtle cues from facial movements, providing real-time insights into the user's emotional state. This technology is especially useful in scenarios where direct verbal feedback might not be possible or reliable. Studies by Zeng et al. (2009) and Barrett et al. (2019) have demonstrated the accuracy of facial analysis in detecting emotions like happiness, sadness, anger, or surprise, which can be correlated with user interactions.

2.2. Multimodal Approaches to UX Evaluation

The integration of multiple biometric methods has become a crucial development in UX research. Tools like the Eyeface system, which combines eye tracking with facial recognition, exemplify this trend. By incorporating different biometric data streams, multimodal tools can provide a richer and more nuanced understanding of the UX (Lasa et al., 2015). Multimodal approaches are particularly useful for capturing emotional data, allowing product developers to assess users' feelings during specific moments of interaction (Akdogan et al., 2018).

Eyeface System and Its Applications

The Eyeface System represents a significant leap forward in emotional UX evaluation. Combining real-time eye-tracking data with facial recognition, allows for a detailed assessment of user engagement and emotional responses. This multimodal system has been used in both gaming and product design, providing actionable insights that traditional UX testing methods might overlook (Lasa et al., 2015). For example, in a gaming context, developers can track which elements of gameplay provoke stress or frustration and make necessary adjustments to improve player satisfaction.

2.3. The Role of Biometric Sensors in Gaming

The gaming industry has been at the forefront of adopting biometric sensors for UX evaluation. The need for continuous improvement in game design and user engagement has led developers to seek more precise methods for evaluating player experiences. As Nacke et al. (2011) demonstrated, combining EEG, eye tracking, and electromyography (EMG) can provide detailed insights into players' interaction with games on psychological and physical levels.

EEG in Gaming

EEG has been instrumental in identifying moments of peak engagement or frustration during gameplay. Players' brainwaves can reveal when they are fully immersed in a game or when they are experiencing cognitive overload due to complex tasks (Awad & Liu, 2018). This data enables developers to tweak game mechanics, level designs, and difficulty curves to optimize the gaming experience (Nacke et al., 2011).

EMG for Physical Response Measurement

EMG, which measures muscle activity, has been used to capture players' physical reactions during gaming sessions. It helps track involuntary responses such as tension in the jaw or hands, providing additional layers of feedback that can enhance game design (Ravaja et al., 2006; Al Eidan, 2013). EMG is particularly useful for evaluating games that require physical interaction, especially those played with motion controllers or virtual reality (VR) headsets.

Eye Tracking in Gaming

Eye-tracking technology is used to assess how players interact with the visual components of a game. By mapping a player's gaze, developers can determine whether game elements, namely, interfaces, tutorials, or visual cues are effective. It can also help identify moments of confusion, where players might struggle to understand the next steps in a game. This insight allows developers to refine game elements to enhance clarity and improve user experience (Duchowski, 2002).

2.4. Applications Beyond Gaming: Education and Product Design

While the gaming industry has been a leader in biometric UX evaluation, other sectors like education and product design are beginning to adopt these technologies. In educational settings, biometric sensors can help assess student engagement and cognitive load during learning activities. By tracking students' eye movements, facial expressions, and brain activity, educators can better understand how different teaching methods impact learning outcomes (Azevedo & Strain, 2011).

Biometrics in Education

Biometric data offers an objective means of evaluating student engagement and emotional responses to educational content. Studies by González-Díez et al. (2024) and Azevedo et al. (2016) have demonstrated how tools like eye tracking and EEG can help educators assess when students are experiencing confusion, frustration, or boredom. This data allows for the real-time adjustment of teaching methods, helping to personalize learning experiences and improve educational outcomes.

Product Design

In product design, biometric sensors offer a valuable tool for understanding users' physical and emotional interaction with a product. By combining data from eye tracking, FEA, and EEG, designers can gain insights into the usability and emotional impact of their products. This multimodal approach allows for a comprehensive evaluation of product effectiveness and user satisfaction (Lasa et al., 2017).

2.5. The Future of Biometric Sensors in UX Evaluation

As biometric technology continues to evolve, its applications in UX evaluation will likely expand into new areas. Advances in ML and artificial intelligence (AI) are already enabling more sophisticated analysis of biometric data, allowing for real-time feedback and adaptive user interfaces, like wearables (Sellahewa et al., 2019). AI-powered systems can analyze biometric data to automatically adjust game difficulty or customize learning experiences based on the user's emotional state (Zeng et al., 2009).

In the future, we can expect to see biometric sensors integrated into more everyday devices, from smartphones to VR headsets, allowing for continuous and unobtrusive UX evaluation across a wide range of industries. This will likely lead to more personalized and adaptive user experiences, further enhancing the role of biometric sensors in the design and development of products and services.

The integration of biometrics with AI and ML heralds a new era of behavior-based recognition, which promises to revolutionize user authentication and fraud prevention (Conley, 2020; Gavrilova et al., 2022). These emerging trends are set to redefine the standards of UX across industries, highlighting the need for ongoing research and development to address challenges related to privacy and data protection.

2.6. Comparison of Traditional Methods and Biometrics in UX Evaluation

Traditional UX evaluation methods, namely, self-reports, observational data, and human-system performance metrics, have been the foundation of understanding user interactions. These approaches present notable limitations. Self-reports depend on users' ability to articulate subjective experiences, which can be influenced by self-awareness biases or reluctance to share true feelings. Observational methods, while offering external perspectives, require analysts to infer psychological states from behaviors, often leading to incomplete assessments. In contrast, biometric technologies provide a transformative alternative by capturing objective, physiological responses that traditional methods often overlook. Tools like eye tracking, GSR, EEG, and facial expression analysis offer real-time, continuous data collection, revealing subconscious reactions and nuanced emotional states that users may not consciously recognize or articulate. This integration of biometric data addresses the gaps in traditional UX evaluation, providing a comprehensive understanding of UXs and enabling the optimization of interfaces and products. Combining biometric data with conventional methods through triangulation enhances the validity and reliability of UX research, creating a robust evaluation framework that capitalizes on the strengths of both approaches.

Biometric technologies are gaining traction in the hospitality industry for enhancing security, improving operational efficiency, and evaluating UX. They can aid in emergency management by tracking individuals and facilitating evacuations (Azevedo & Strain, 2011). Various biometric authentication methods are being explored for secure guest services and hospitality operations (Warren, 2010). Recent studies have demonstrated the potential of biometric methods to complement traditional UX research approaches, offering real-time, objective data on users' neurophysiological responses to design elements and interactive experiences (Aloi et al., 2024). The integration of biometrics in hospitality and tourism is expected to regulate internal processes and improve services, with applications ranging from access control to personalized guest experiences (Kang et al., 2007).

Traditional UX evaluation methodologies face critical limitations in capturing comprehensive UXs:

Subjectivity and Bias

Self-reports (e.g., surveys, interviews) rely on users' ability to articulate their experiences. These are prone to biases like social desirability or recall errors and fail to capture subconscious or nuanced reactions.

Incomplete Data

Observational methods depend on human interpretation of behavior, which may not accurately infer users' psychological or emotional states.

Conventional metrics (e.g., click-through rates) are limited to performance measures and neglect emotional or cognitive dimensions.

Granularity

Traditional methods often provide static, point-in-time insights, overlooking the continuous and dynamic nature of UXs.

Inefficiency in Capturing Emotional and Cognitive States

These methods fail to quantify subtle affective or cognitive responses, namely, micro-expressions or cognitive load, which are integral to understanding user satisfaction.

Limited Applicability to Advanced Interfaces

Emerging environments like AR/VR or AI-driven systems demand real-time, adaptive evaluation mechanisms, which traditional approaches are ill-equipped to address. Biometric methods offer promising avenues for enhancing UX evaluation in various domains, including healthcare (Cheng & Lin, 2023). These techniques can complement traditional UX assessment methods by providing objective, real-time data on users' neurophysiological responses (Aloi et al., 2024). Biometric sensors, eye tracking and EEG, can capture hidden psychological states and affect, potentially offering richer insights into UX than self-report measures alone (Kula et al., 2019). Studies have demonstrated the feasibility of using brain biometric signals to identify emotions in older adults, contributing to UX evaluation models (Iriarte et al., 2021). In healthcare applications, biometric features are increasingly used for user identification and authentication. Efficient feature evaluation and selection methods, like the Farness value approach, can improve the accuracy and efficiency of biometric-based classifiers for user identification (Kayastha & Sha, 2019). While promising, researchers advocate for establishing ethical guidelines and standardization strategies for conducting biometric user research (Aloi et al., 2024).

3. Methodology

For the chapter titled "Biometrics as a key factor for emergent User Experience Evaluation", the research methodology combines a scoping review and case study approach. This mixed-method strategy allows a comprehensive exploration of how biometrics enhance UX, ensuring breadth and depth of analysis. The scoping review methodology is selected to map out existing research on biometrics in UX evaluation and identify gaps in knowledge (Eisenhardt, 1989; Flyvbjerg, 2006; Yin, 2018). Unlike systematic reviews, scoping reviews are more flexible and exploratory, allowing the researcher to cover broad topics where different methodologies or approaches might be applied. The following steps will be taken in conducting the scoping review:

1. Research Questions

The main research question guiding the scoping review is: *How have biometric technologies been applied to emergent UX evaluation across various industries, and what are the gaps in current methodologies?* Sub-questions include:

- What types of biometric data (e.g., heart rate, eye-tracking, GSR are most frequently employed in UX research?

- How do biometric tools compare with traditional UX evaluation techniques?
- What are the challenges and limitations of using biometrics in emergent digital environments (e.g., VR/AR, AI-driven interfaces)?

2. Search Strategy
A systematic search is conducted in academic databases including IEEE Xplore, Scopus, Web of Science, and Google Scholar. Keywords include '*biometrics*', "*user experience evaluation*", '*UX*', "*biometric data*", "*eye-tracking*", "*human-computer interaction*", "*AR/VR UX evaluation*", and "*wearable technology in UX*". This broad keyword selection ensures the inclusion of both foundational and emergent technologies in UX evaluation.

3. Study Selection Criteria
Inclusion criteria are:

- Peer-reviewed journal articles, conference papers, and industry white papers published between 2005 and 2024.
- Studies focusing on the application of biometric tools in UX evaluations.
- Studies across sectors like healthcare, education, entertainment, and banking.

Exclusion criteria are:

- Articles not focused on biometrics for UX.
- Papers older than 2005, unless seminal or highly cited.

The scoping review includes a thematic analysis of the extracted data. Key themes like types of biometric tools, user engagement, accuracy of UX assessment, and ethical considerations are analyzed to provide a conceptual framework for understanding how biometric tools are transforming UX evaluation.

To complement the broad findings from the scoping review, a case study approach delves into specific applications of biometrics in real-world UX evaluations (Baxter & Jack, 2008; Stake, 1995). The case study method will provide a contextual, detailed, and in-depth understanding of how biometric data is used to enhance UX across different settings, analyzing the case of Spanish startup Gestualy.

The combination of a scoping review and case study methodology allows for a well-rounded understanding of how biometrics influence UX evaluation. The scoping review offers a broad perspective by synthesizing existing literature, while the case studies provide tangible examples of how biometrics are applied in practice. This mixed-method approach is crucial to understanding both the theoretical and practical implications of biometric tools in UX evaluation. This methodology provides a clear path for understanding the role of biometrics in UX, offering broad insights and detailed examples.

4. Case study

A case study of Gestualy, a leading startup in biometric solutions, illustrates the transformative impact of these technologies. Gestualy integrates advanced gesture and motion recognition to create intuitive and engaging user interfaces. By capturing and analyzing users' gestures, the platform offers real-time feedback and customization, significantly enhancing the UX. This innovative approach is particularly beneficial in settings requiring quick, accurate identification without physical contact, like healthcare facilities and customer service centers.

4.1. Use of Gestualy in Hotels

The hospitality industry is highly competitive, requiring businesses to constantly improve customer service while optimizing operational efficiency. Gestualy, an advanced biometric and gesture-based system, offers a unique solution by transforming how hotels manage guest interactions and security. This case study examines two critical applications of the Gestualy system in hotels: enhancing guest experiences through personalized services and improving security by automating check-ins and access controls.

The ability to provide tailored services is a key differentiator in the hospitality sector. Gestualy facilitates this by integrating biometric recognition with AI, enabling hotels to deliver highly personalized experiences for each guest. Upon arrival, guests can be identified through facial recognition or other biometric inputs, allowing hotel staff to access detailed information about their preferences, past stays, and special requests instantly. This feature streamlines check-in procedures, making the process faster and more convenient.

For example, when a guest enters the hotel lobby, the system automatically identifies them and prompts staff with tailored information. This level of personalization enhances customer satisfaction, creating a memorable and seamless experience. Additionally, it eliminates the need for repetitive tasks by guests, like filling out forms or verifying identities multiple times, which can be a point of frustration.

Moreover, Gestualy can integrate with room management systems to automatically adjust room settings—like temperature, lighting, and entertainment systems—based on guest preferences. This integration enhances the guest's comfort without requiring manual adjustments. In luxury or high-end hotels, such levels of personalization are becoming standard expectations, and systems like Gestualy are helping hoteliers exceed these expectations with greater efficiency.

One of the most significant challenges in the hotel industry is ensuring robust security without compromising guest comfort. Gestualy's biometric features offer a solution by automating access control. Once a guest has been biometrically enrolled in the system, their identity can be seamlessly authenticated at various access points, namely, room doors, elevators, and exclusive areas like the gym or swimming pool.

Besides enhancing security, it reduces the administrative burden on hotel staff. By eliminating the need for physical keys or cards, hotels can significantly reduce the risk of lost or stolen access credentials. Gestualy's data encryption and secure communication protocols ensure that guest information remains private and protected from unauthorized access.

Additionally, the system can be used for visitor management, allowing hotels to better control who has access to certain areas of the property. This is especially important in large hotels or resorts where controlling guest movement is critical to maintaining security. For example, a guest enrolled in the system could gain access to specific floors or areas, while others would be restricted based on their level of access.

A high-end hotel in Barcelona implemented the Gestualy system across its operations to provide an elevated guest experience while enhancing security. Upon implementation, the hotel noted a 30% reduction in check-in times and a 20% increase in positive customer feedback related to personalized service. Guests appreciated the seamless check-in experience, and many lauded the personalized room settings available upon arrival. Moreover, the system enabled real-time access control, significantly reducing the occurrence of lost room keys.

The hotel also reported increased efficiency in its housekeeping and maintenance departments, as Gestualy enabled these staff members to access real-time information about room occupancy and service requests. This allowed the hotel to streamline room servicing, ensuring that guest needs were met without unnecessary delays or disturbances.

Gestualy offers hotels a dual advantage: it enhances the guest experience and improves operational efficiency through advanced biometric access control. The case study of a high-end hotel in Barcelona demonstrates the system's ability to reduce operational inefficiencies while significantly improving customer satisfaction. As hotels continue to seek innovative ways to differentiate themselves, systems like Gestualy are becoming an essential component of modern hospitality management.

4.2. Use of Gestualy in Healthcare

The healthcare sector faces unique challenges in balancing patient care with operational efficiency. One critical issue is the secure management of patient data and access control to sensitive areas. Gestualy, a biometric-based system offers a solution that enhances both security and patient care by integrating gesture and facial recognition technologies with healthcare operations. This case study focuses on two primary applications of the Gestualy system in healthcare: improving patient identification and automating access control in medical facilities.

In healthcare, accurate patient identification is critical to ensuring that the right treatments are administered to the right patients. Gestualy leverages biometric recognition—such as facial recognition, fingerprint scanning, or gesture-based interactions—to automate patient identification at every point of contact within the healthcare facility. It eliminates the risk of patient misidentification, a frequent source of medical errors.

When a patient arrives at a hospital or clinic, the Gestualy system can automatically identify them, pulling up their medical history, allergy information, and ongoing treatments. This streamlines the check-in process and ensures that healthcare providers have instant access to critical patient data. The system can alert medical staff to specific health conditions or requirements, including allergies to medications, which need immediate attention.

Gestualy can be integrated with electronic health record (EHR) systems to facilitate real-time updates. As patients undergo treatments or consultations, their biometric information can be used to authenticate their identity, ensuring that their medical records are updated accurately and securely. This minimizes the likelihood of human error in record-keeping, improving overall patient safety.

Access control is another area where Gestualy proves valuable in healthcare settings. Hospitals and clinics must restrict access to sensitive areas like operating rooms, medication storage, and patient wards. Traditional keycard systems or manual check-ins can be cumbersome and prone to breaches. Gestualy, through its biometric authentication capabilities, offers a more secure and efficient solution. Only authorized personnel would be allowed access to an operating room, based on their biometric profile stored in the Gestualy system. This ensures that only certified and permitted individuals can enter, enhancing both security and patient safety. The system can also track who enters and exits these sensitive areas, providing a real-time log that can be audited for compliance with healthcare regulations.

In high-risk areas like intensive care units (ICUs) or neonatal wards, where access must be tightly controlled, Gestualy's ability to automate entry based on biometric data adds an additional layer of protection. It ensures that only the appropriate staff members, namely, doctors or nurses assigned to specific patients, have access to critical areas, thus reducing the risk of contamination or unauthorized intervention.

A large hospital in Madrid implemented the Gestualy system to improve both patient care and facility security. Prior to Gestualy, the hospital faced challenges related to patient misidentification and unauthorized access to restricted areas. Since the implementation, the hospital reported a 40% reduction in administrative errors related to patient identification and a 25% improvement in the security of high-risk areas like the ICU and operating rooms. The hospital also noted improvements in operational efficiency, as healthcare staff no longer needed to manually check access credentials or verify patient identities multiple times throughout the day. With Gestualy, these processes became automatic, freeing up staff to focus more on patient care rather than administrative tasks. Additionally, the integration with the hospital's EHR system enabled seamless updates to patient records in real time, improving the accuracy of medical documentation and reducing the risk of errors during patient handovers between shifts.

The case study of a large hospital in Madrid highlights how Gestualy's integration into healthcare environments can significantly improve the security of medical facilities and the quality of patient care. By automating patient identification and access control, the system reduces the potential for errors and improves operational efficiency. As healthcare facilities continue to prioritize patient safety and data security, systems like Gestualy are becoming indispensable tools in modern medical management.

Retailers utilize facial recognition to personalize shopping experiences and streamline payment processes, leading to higher customer satisfaction and loyalty. Similarly, in customer service, biometrics reduce wait times and improve service efficiency by enabling quick and secure access to user accounts.

As a general framework of biometrics approach to bridging the gap in comparison with traditional methods, particularly within the Gestualy technology range of application. Biometric technologies provide unique solutions to these shortcomings by leveraging physiological and behavioral data:

Objective, Real-time Data

Biometric tools like eye-tracking, EEG, and GSR deliver continuous, real-time insights into users' cognitive and emotional states.

These methods objectively capture subconscious reactions and physiological changes that users might not report or even be aware of. Real-time insights from biometric tools like EEG and GSR can identify pain points during user interactions. This data allows designers to streamline navigation paths and optimize user flows, resulting in a more intuitive UX.

Granular Analysis

Techniques such as EEG measure brain activity, allowing detailed analysis of cognitive load and emotional engagement throughout the user journey. Triangulating biometric data with traditional methods (e.g., surveys, interviews) strengthens UX evaluation frameworks. Eye-tracking data, for instance, can validate user-reported attention patterns, making insights more robust and actionable.

Enhanced Emotional and Behavioural Understanding

Tools like facial expression analysis detect micro-expressions and emotional nuances, providing a richer understanding of user responses to interfaces. Combining biometric data, such as eye-tracking and facial expression analysis, with traditional UX metrics can dynamically adapt interfaces to user preferences and emotional states. For example, a website could alter its layout in real time to reduce cognitive load for frustrated users, enhancing usability. Implicit biometric authentication, including behavioral and physiological monitoring, can maintain security without disrupting user experiences. For instance, multi-biometric systems in online banking can ensure seamless yet secure transactions.

Application in Complex Environments

Biometrics enable usability testing in dynamic settings and immersive environments, like VR and AR. This is essential for evaluating user engagement and satisfaction in adaptive systems driven by AI. By tracking physiological responses, designers can adjust virtual environments in real time to improve comfort and engagement, addressing challenges such as motion sickness or cognitive overload.

Integration with Traditional Methods

Triangulating biometric data with traditional methods, e.g., surveys, etc., strengthens UX evaluation frameworks. Eye-tracking data can validate user-reported attention patterns, making insights more robust and actionable.

Biometric technologies thus transform UX evaluation by offering deeper, more accurate insights into user behavior and satisfaction, paving the way for adaptive, user-centered designs in modern interfaces. This innovation addresses the core limitations of traditional methodologies, enabling the development of seamless and emotionally resonant UXs across industries.

5. Conclusions

The research presented herein highlights the critical role biometrics play in advancing UX evaluation, particularly in the context of emerging technologies and interfaces. By leveraging biometric technologies, like s eye tracking, facial expression analysis, and EEG, UX researchers can gain unprecedented insight into the cognitive and emotional states of users, far beyond what traditional methods can provide. The primary advantage of biometrics in UX evaluation lies in its ability to capture real-time, objective data, offering a holistic understanding of how users interact with products, services, and digital environments.

A key conclusion drawn from this study is that biometric tools provide a more granular, continuous stream of data, which captures subconscious reactions that users themselves may be unaware of. This real-time insight is invaluable for identifying specific design pain points, moments of confusion, or emotional engagement that traditional self-reporting methods often overlook. The data richness offered by biometrics enables the identification of cognitive load, emotional responses, and micro-expressions that influence user satisfaction and performance. This granularity paves the way for more personalized and adaptive UX design, particularly in complex environments like VR, AR, and AI-driven interfaces.

Practically, biometric technologies have been demonstrated as game-changers across industries like healthcare, hospitality, and retail, as illustrated by the Gestualy case study. In the hospitality industry, biometric solutions including gesture-based systems streamline operational efficiency and enhance personalized customer experiences. In healthcare, these technologies ensure more accurate patient identification and secure access control, significantly reducing medical errors and improving operational workflows. These case studies underline the transformative potential of biometrics in both enhancing UX and improving security across sectors.

In educational settings, biometric tools like eye-tracking and EEG can monitor student engagement and cognitive load during lessons. This real-time data enables teachers to identify when students are struggling with concepts or losing focus. An adaptive learning platform could use this data to adjust the difficulty level of the material or recommend supplementary resources tailored to the student's current needs, thereby enhancing personalized education outcomes.

Retailers can utilize facial recognition to identify returning customers and access their shopping history or preferences. Integrated with in-store displays, biometric data can guide customers toward products they are more likely to purchase. A digital signage system in a clothing store can analyze customers' facial expressions to assess satisfaction with a displayed product and adjust recommendations in real-time, optimizing the shopping experience and increasing sales.

In airports and public transit systems, biometric authentication (e.g., fingerprint or iris scanning) can streamline identity verification processes, reducing bottlenecks and improving passenger flow. A smart boarding system at an airport could use facial recognition to match passengers with their boarding passes automatically, significantly reducing wait times while maintaining security.

This chapter explores how the integration of biometrics with AI and ML holds the potential to revolutionize UX further. AI-driven systems can use biometric data to create dynamic, responsive systems that adapt in real-time to users' emotional and cognitive states. This capability is particularly beneficial in creating seamless, personalized experiences in industries that require high levels of security and usability, like banking, healthcare, and customer service (Tatineni et al., 2022).

The adoption of biometrics in UX evaluation offers a transformative approach to understanding user interaction, satisfaction, and performance. The data provided by biometric systems extends far beyond what traditional UX methods can capture, offering richer insights into user behavior, preferences, and emotions. As biometric technology continues to evolve, its integration with AI and ML promises to drive even more personalized and adaptive UXsDespite its many advantages, the adoption of biometric tools in UX evaluation faces challenges.

6. Research Limitations

While the integration of biometric technologies in UX evaluation is promising, this study has revealed several limitations that need to be addressed to maximize their effectiveness. First, the implementation of biometric tools requires substantial investment in both technology and expertise. They often require complex setups and calibration, which can be costly and time-consuming, particularly for smaller organizations with limited resources. This may limit the widespread adoption of these tools, especially in industries where budget constraints are a significant factor.

Another limitation is the issue of data quality and reliability. Biometric systems rely heavily on the quality of sensed data, and external factors like lighting, user movement, and the physical setup of sensors can impact the accuracy of the data collected. Facial recognition and eye-tracking technologies can be sensitive to variations in lighting, while EEG and GSR measurements may be affected by user movement or environmental conditions. These inconsistencies can lead to skewed data, making it difficult to draw reliable conclusions about UX.

User acceptance is a critical limitation in the application of biometrics. Many users express concerns over privacy and the security of their biometric data, especially in industries where sensitive personal information is involved, such as healthcare and banking. The ethical implications of biometric data collection and storage cannot be overlooked, as the potential for misuse or unauthorized access poses significant risks. This limitation underscores the need for clear ethical guidelines, and transparent policies for how biometric data is collected, stored, and used.

Moreover, there is a risk that users may feel uncomfortable or self-conscious when their biometric data is being collected, which could influence their natural behavior, leading to biased results. The awareness of being monitored through biometric sensors could inadvertently cause users to alter their interactions, thus affecting the authenticity of the data collected. This limitation highlights the need for further research into creating less intrusive biometric systems that can capture data without disrupting the user's natural experience.

Lastly, the lack of standardization in biometric UX evaluation presents a challenge for researchers and practitioners. The field is still in its developmental stage, and there is a lack of widely accepted frameworks or benchmarks for evaluating biometric data in UX studies, making it difficult to compare findings across different studies and industries, hindering the development of a comprehensive understanding of how biometric data can be effectively used to enhance UX.

Hence, while biometrics offer immense potential for improving UX evaluation, their adoption is not without challenges. Issues related to cost, data reliability, user acceptance, and the need for ethical and standardization frameworks must be addressed to ensure that biometric technologies can be effectively integrated into UX research. Future research should focus on overcoming these limitations by exploring more cost-effective, scalable, and user-friendly biometric solutions, and developing standardized methodologies for biometric data collection and analysis in UX evaluation.

References

Abdulkareem, A.M., & Gordon, A. (2023). Evaluating the usability and user acceptance of biometric authentication in different applications. *Quarterly Journal of Emerging Technologies and Innovations*, 8(2), 1–10.

Akdogan, D., Karaoglan Altop, D., Eskandarian, L., & Levi, A. (2018). Secure key agreement protocols: Pure biometrics and cancelable biometrics. *Computer Networks*, 142, 33-48. https://doi.org/10.1016/j.comnet.2018.06.001

Al Eidan, R.M. (2013). Hand biometrics: Overview and user perception survey. In: *2013 Second International Conference on Informatics & Applications (ICIA)*, (pp. 252–257). https://doi.org/10.1109/icoia.2013.6650265.

Aloi, D., Bouzit, S., & Li, J. (2024). Biometric methods for user research: Three case studies. In: *Extended Abstracts of the CHI Conference on Human Factors in Computing Systems* (pp. 1–8). Honolulu, Hawai, US.

Awad, A., & Liu, Y. (2018). Cognitive biometrics for user authentication. In: Obaidat, M., Traore, I., Woungang, I. (Eds.), *Biometric-Based Physical and Cybersecurity Systems* (pp. 387–399). Springer. https://doi.org/10.1007/978-3-319-98734-7_15

Azevedo, R., & Strain, A.C. (2011). Integrating cognitive, metacognitive, and affective regulatory processes with MetaTutor. In: *New Perspectives on Affect and Learning Technologies* (pp. 141–154). Springer.

Azevedo, A.I., Santos, H.D., Sá, V.J., & Lopes, N.V. (2016). Behavioural biometrics for authentication and stress detection – A case study with children. *Global Security, Safety and Sustainability*, 266–280. 11th International Conference on Global Security, Safety, and Sustainability, ICGS3 2017. https://doi.org/10.1007/978-3-319-51064-4_21

Barrett, L.F., Adolphs, R., Marsella, S., Martinez, A.M., & Pollak, S.D. (2019). Emotional expressions reconsidered: Challenges to inferring emotion from human facial movements. *Psychological Science in the Public Interest*, 20(1), 1–68.

Baxter, P., & Jack, S. (2008). Qualitative case study methodology: Study design and implementation for novice researchers. *The Qualitative Report*, 13(4), 544-559. https://doi.org/10.46743/2160-3715/2008.1573.

Blanco-Gonzalo, R., Sanchez-Reillo, R., Miguel-Hurtado, O., & Bella-Pulgarin, E. (2014). Automatic usability and stress analysis in mobile biometrics. *Image and Vision Computing*, 32(12), 1173-1180. https://doi.org/10.1016/j.imavis.2014.09.003.

Brockly, M., Elliott, S.J., Guest, R.M., & Blanco-Gonzalo, R. (2015). Human-biometric sensor interaction. In: Li, S., & Jain, A. (Eds.), *Encyclopedia of Biometrics*. Springer. https://doi.org/10.1007/978-3-642-27733-7_2261-3.

Brostoff, G. (2017). Adoption problems? How UX could boost biometrics. *Biometric Technology Today*, 2017(9), 9–11.

Cheng, C. & Lin, C. (2023). Building a low-cost wireless biofeedback solution: Applying design science research methodology. *Sensors*, 23(6), 2920. https://doi.org/10.3390/s23062920.

Conley, Q. (2020). Learning user experience design (LUX): Adding the 'L' to UX research using biometric sensors. In: *Research Methods in Learning Design and Technology* (pp. 146–160). Routledge.

Corsetti, B., Sanchez-Reillo, R., & Guest, R.M. (2020). Ergonomics in mobile fingerprint recognition systems: A user interaction evaluation. In: *Advances in Human Factors and Systems Interaction: Proceedings of the AHFE 2020 Virtual Conference on Human Factors and Systems Interaction* (pp. 382–389). Springer.

Cowley, B., Filetti, M., Lukander, K., Torniainen, J., Henelius, A., Ahonen, L., ... & Jacucci, G. (2016). The psychophysiology primer: A guide to methods and a broad review with a focus on human–computer interaction. *Foundations and Trends in Human–Computer Interaction*, 9(3–4), 151–308.

Duchowski, A.T. (2002). A breadth-first survey of eye-tracking applications. *Behavior Research Methods, Instruments, & Computers*, 34(4), 455–470.

Eisenhardt, K.M. (1989). Building theories from case study research. *Academy of Management Review*, 14(4), 532–550. https://doi.org/10.5465/amr.1989.4308385.

Ekman, P., & Friesen, W.V. (1969). The repertoire of nonverbal behavior: Categories, origins, usage, and coding. *Semiotica*, 1(1), 49–98.

El-Abed, M., Giot, R., Hemery, B., & Rosenberger, C. (2012). Evaluation of biometric systems: A study of users' acceptance and satisfaction. *International Journal of Biometrics*, 4(3), 265–290. https://doi.org/10.1504/IJBM.2012.047644.

Elliott, S.J., & Kukula, E.P. (2010). A definitional framework for the human/biometric sensor interaction model. In: *Biometric Technology for Human Identification VII*, 7667, 164–171. SPIE.

Fernandez-Saavedra, B., Alonso-Moreno, R., Uriarte-Antonio, J., & Sanchez-Reillo, R. (2010). Evaluation methodology for analyzing usability factors in biometrics. *IEEE Aerospace and Electronic Systems Magazine*. https://doi.org/10.1109/MAES.2010.5552609.

Flyvbjerg, B. (2006). Five misunderstandings about case-study research. *Qualitative Inquiry*, 12(2), 219–245. https://doi.org/10.1177/1077800405284363.

Gavrilova, M., Luchak, I., Sudhakar, T., & Tumpa, S.N. (2022). Artificial intelligence in biometrics: Uncovering intricacies of human body and mind. In: *Advances in Selected Artificial Intelligence Areas: World Outstanding Women in Artificial Intelligence* (pp. 123–169). Springer.

González-Díez, I., Varela, C., & Sáiz-Manzanares, M.C. (2024). The use of integrated multichannel records in learning studies in higher education: A systematic review of the last 10 years. *Computers*, 13(4), 96.

Gudala, L., Reddy, A.K., Sadhu, A.K.R., & Venkataramanan, S. (2022). Leveraging biometric authentication and blockchain technology for enhanced security in identity and access management systems. *Journal of Artificial Intelligence Research*, 2(2), 21–50.

Hassenzahl, M., & Tractinsky, N. (2006). User experience: A research agenda. *Behaviour & Information Technology*, 25(2), 91–97.

Huseynov, S., Kassas, B., Segovia, M.S., & Palma, M.A. (2019). Incorporating biometric data in models of consumer choice. *Applied Economics*, 51(14), 1514–1531.

Idoko, I. (2024). Big data and AI in employment: The dual challenge of workforce replacement and protecting customer privacy in biometric data usage. *Global Journal of Engineering and Technology Advances*, 19(2), 089–106. https://doi.org/10.30574/gjeta.2024.19.2.0080.

Iriarte, A.A., Erle, G.L., Etxabe, M.M., Apraiz-Iriarte, A., Lasa-Erle, G., & Etxabe,, M. (2021). Evaluating user experience with physiological monitoring: A systematic literature review. *Dyna New Technol*, 8(1).

Jick, T.D. (1979). Mixing qualitative and quantitative methods: Triangulation in action. *Administrative Science Quarterly*, 24(4), 602–611.

Kang, J.S. (2010). Mobile iris recognition systems: An emerging biometric technology. *Procedia Computer Science*, 1(1), 475–484.

Kloet, M. and Yang, S. (2022). The effects of anthropomorphism and multimodal biometric authentication on the user experience of voice intelligence. *Frontiers in Artificial Intelligence*, 5. https://doi.org/10.3389/frai.2022.831046.

Koonsanit, K., & Nishiuchi, N. (2020). Classification of user satisfaction using facial expression recognition and machine learning. In: *2020 IEEE Region 10 Conference (TENCON)*, 561–566.

Kukula, E., Sutton, M., & Elliott, S. (2010). The human–biometric-sensor interaction evaluation method: biometric performance and usability measurements. *IEEE Transactions on Instrumentation and Measurement*, 59, 784–791. https://doi.org/10.1109/TIM.2009.2037878.

Kula, I., Branaghan, R.J., Atkinson, R.K., & Roscoe, R.D. (2019). Assessing user experience via biometric sensor affect detection. In: *Human Performance Technology: Concepts, Methodologies, Tools, and Applications* (pp. 698–714). IGI Global.

Kumar, A., & Zhang, D. (2010). Improving biometric authentication performance from the user quality. *IEEE Transactions on Instrumentation and Measurement*, 59, 730–735. https://doi.org/10.1109/TIM.2009.2028773.

Lallemand, C., & Gronier, G. (2019). *Measuring User Experience: A Pragmatic Guide to UX Evaluation*. Springer.

Lasa, G., Justel, D., & Retegi, A. (2015). Eyeface: A new multimethod tool to evaluate the perception of conceptual user experiences. *Computers in Human Behavior*, 52, 359–363.

Lasa, G., Justel, D., Gonzalez, I., Iriarte, I., & Val, E. (2017). Next generation of tools for industry to evaluate the user emotional perception: The biometric-based multimethod tools. *The Design Journal*, 20(sup 1), S2771–S2777.

Li, X., Wang, Z., Chen, C., & Niu, J. (2013). Applying biometrics to design a three-factor remote user authentication scheme with key agreement. *Security and Communication Networks*, 7(10), 1488–1497. https://doi.org/10.1002/sec.767.

Mauss, I.B., & Robinson, M.D. (2009). Measures of emotion: A review. *Cognition and Emotion*, 23(2), 209–237.

Nacke, L. (2009). Next generation testing: Biometric analysis of player experience. In: *Game Developers Conference*. Canada.

Nacke, L.E., Kalyn, M., Lough, C., & Mandryk, R.L. (2011, May). Biofeedback game design: Using direct and indirect physiological control to enhance game interaction. In: *Proceedings of the SIGCHI Conference on Human Factors in Computing Systems* (pp. 103–112). Vancouver BC, Canada.

Picard, R.W. (1997). *Affective Computing*. MIT Press.

Poole, A., & Ball, L.J. (2006). Eye tracking in human-computer interaction and usability research: Current status and future prospects. *Encyclopedia of Human Computer Interaction.* Idea Group Reference.

Ravaja, N., Saari, T., Turpeinen, M., Laarni, J., Salminen, M., & Kivikangas, M. (2006). Spatial presence and emotions during video game playing: Does it matter with whom you play? *Presence: Teleoperators and Virtual Environments*, 15(4), 381–392.

Rengarajan, A., & Rahul, C.R. (2024). Behavioural biometrics as a user authentication mechanism in ISMS. *International Journal of Innovative Research in Computer and Communication Engineering*, 12(03), 1699–1706. https://doi.org/10.15680/ijircce.2024.1203054.

Ryu, R., Yeom, S., Herbert, D., & Dermoudy, J. (2023). A comprehensive survey of context-aware continuous implicit authentication in online learning environments. *IEEE Access*, 11, 24561–24573. https://doi.org/10.1109/access.2023.3253484.

Salminen, M., Kivikangas, J.M., Ravaja, N., & Kallinen, K. (2009). Frontal EEG asymmetry in the study of player experiences during competitive and cooperative dual play. *Proceedings of Game and Entertainment Technologies* (pp. 44–55).

Sarier, N.D. (2022). Privacy preserving biometric authentication on the blockchain for smart healthcare. *Pervasive and Mobile Computing*, 86, 101683.

Sarkar, A., & Singh, B.K. (2020). A review on performance, security, and various biometric template protection schemes for biometric authentication systems. *Multimedia Tools and Applications*, 79(37), 27721–27776.

Sellahewa, H., Ibrahim, N., & Zeadally, S. (2019). Biometric authentication for wearables. *Biometric-Based Physical and Cybersecurity Systems*, 355–386.

Schiavone, E., Ceccarelli, A., Carvalho, A., & Bondavalli, A. (2019). Design, implementation, and assessment of a usable multi-biometric continuous authentication system. *International Journal of Critical Computer-Based Systems*, 9(3), 215. https://doi.org/10.1504/ijccbs.2019.104490.

Stake, R.E. (1995). *The Art of Case Study Research.* SAGE Publications.

Tatineni, S. (2022). Customer authentication in mobile banking-MLOps practices and AI-driven biometric authentication systems. *Journal of Economics & Management Research*, 3(4), 1–5.

Theofanos, M., Stanton, B., Micheals, R., & Orandi, S. (2007). Biometric systematic uncertainty and the user. *First IEEE International Conference on Biometrics: Theory, Applications, and Systems* (pp. 1–6). https://doi.org/10.1109/BTAS.2007.4401918.

Toutara, F., & Spathoulas, G. (2020). A distributed biometric authentication scheme based on blockchain. In: *IEEE International Conference on Blockchain* (pp. 470–475). doi: 10.1109/Blockchain50366.2020.00068.

Warren, K. (2010). Weighing the option of biometrics in the hospitality industry. *Worldwide Hospitality and Tourism Themes*, 2(1), 100–109.

Yamazaki, T. (2019). Case studies on biometric application for quality-of-experience evaluation in communication. *Advances in Biometrics: Modern Methods and Implementation Strategies*, 227–247. https://doi.org/10.1007/978-3-030-30436-2_11.

Yin, R.K. (2018). *Case Study Research and Applications: Design and Methods*. SAGE Publications.

Zeng, Z., Pantic, M., Roisman, G.I., & Huang, T.S. (2009). A survey of affect recognition methods: Audio, visual, and spontaneous expressions. *IEEE Trans Pattern Anal Mach Intell.*, 31(1), 39–58.

Zhang, D., Liang, Z., Song, F., & Xu, Y. (2009). *From Single Biometrics to Multi-Biometrics* (pp. 254–272). IGI Global. https://doi.org/10.4018/978-1-60566-200-8.ch012.

3

Application and Development of Emotional and Attentional Tools: A Face Reading, Eye Tracking, and Electrodermal Activity Approach

Alvaro Saavedra[1(0000-0002-7629-2252)], Ana Moreno-Lobato[2*(0000-0002-0114-0975)] and Aroa Costa-Feito[3(0000-0001-5502-346X)]

[1] Institute for Advanced Research in Business and Economics (INARBE), Public University of Navarre, Pamplona, Spain
[2] Universidad de Extremadura
[3] University of León

1. Introduction

The premise for the knowledge of implicit responses is that the nervous system (NS) is altered automatically and simultaneously reacting to stimuli (Kreibig, 2010; Levenson, 2014). Therefore, the study of implicit emotional reactions is carried out through the analysis and evaluation of the autonomic nervous system (ANS). The ANS is one of the key elements of behavior through the regulation of emotional reactions (Vila, 1996), acting as bottom-up and antecedents of cognitive functions (Levenson, 2014).

Technically, the NS is a set of structures that conducts signals throughout the body to adapt the various systems of the organism to external and internal changes (Levenson, 2014). Several criteria can be used to categorize the signals and how they function. According to the

*Corresponding author: anamorenol@unex.es

anatomical position of the reactions, a distinction is made between the central nervous system, consisting of the brain and the spinal; and the peripheral nervous system, composed of the nerves branching off from the spinal cord and extending to all parts of the body cord (Al-Nafjan et al., 2023). The latter is divided into somatic and ANSs, the former being responsible for voluntary reactions and movements of the body, and the latter for implicit and involuntary reactions (Kreibig, 2010). Within the autonomic peripheral nervous system, two types of nervous reactions are differentiated, the sympathetic and the parasympathetic, the former being the one that reacts to alarming or relevant stimuli and the parasympathetic to rest and calm (Kreibig, 2010; Li et al., 2015).

Behavior consists of explicit and implicit responses, also related to cognitive and affective outcomes. This is the main reason that explains the inclusion of ANS in understanding the behavioral process. Through neuroscience, psychology, and psychophysiology, it is possible to evaluate various implicit responses of great relevance in people's evaluation and decision-making processes (Vila, 1996). Consumer neuroscience focuses on the analysis and study of the underlying neural conditions and processes that take place during consumption and decision-making processes through a psychological and behavioral interpretation of the results. It is understood as an interdisciplinary area that takes methodologies and tools from neuroscience to understand the biological foundations of behavior (Casado-Aranda & Sánchez-Fernández, 2022).

Neuroscience approaches measurements of the NS through the application of tools from neurology, psychology, and psychophysiology (Kreibig, 2010; Lin et al., 2024). Within consumer neuroscience, two methodological techniques have been developed: neuroimaging and biometry. Marketing applies psychophysiology through biometric techniques with the aim of understanding the emotional and attentional processes. Psychophysiology is a branch of psychology based on behavior and is largely related to physiology as it is a science that evaluates the functioning of various human systems by analyzing organisms through their psychological interpretation (Vila, 1996). Therefore, consumer psychophysiology is the science that evaluates physiological changes through the psychological and behavioral interpretation of these changes (Lajante & Lux, 2020; Vila, 1996).

The measurement tools used are drawn from physiology, as purely emotional and attentional physical changes are assessed in the moment, through the evaluation of the somatic nervous system, which serves as an indicator of psychological change (Li, 2019). The relationship between emotionality and physiological changes in the NS is an issue on which authors do not yet have a consensus (Lajante & Lux, 2020).

The benefit of the application of these biometric methodological tools is that they allow real emotional measurements to be made, not of the memory or interpretation of these. Additionally, they can develop the assessment of changes at various points in the consumption process and reduce interpretation or confirmation biases (Li et al., 2015). Several tools have been applied in the study of ANS reactions: hormonal, involuntary movements, and psychophysiological data (Kreibig, 2010; Levenson, 2014; Li et al., 2015), but trends in marketing and previous studies support the relevance of the application of biometric tools as reliable and valid indicators of emotionality and attention (Kreibig, 2010; Li, 2019).

The chapter aims to analyze three biometric assessment tools of great interest: facial reading, eye tracking, and electrodermal activity. These tools have represented a breakthrough in the consumer research approaches, and their study, analysis, and application represent a challenging research line at a methodological level.

2. Facial Reading

2.1. Overview of Facial Reading Technologies

The human face can be seen as a map that reveals a wealth of information about our emotions, intentions, and moods, making it a crucial element for social interaction and non-verbal communication. Among the various biometric techniques, facial recognition and face reader stand out as those of significant interest for recognizing different aspects of individuals (Kaur et al., 2020). Within the fields of Artificial Intelligence (AI) and machine learning (ML), facial recognition systems are employed to verify individuals' identities (Zhou et al., 2023). Specifically, facial recognition involves capturing human facial features from images or videos, automatically detecting and tracking a face, and authenticating users' identities (Shiau et al., 2023), whereas face reading is a somatic biometric method that analyzes facial characteristics to identify physiological aspects (Berčík et al., 2024). It focuses on examining facial expressions, capturing the movement of the eyebrows, eyes, mouth, nose, and forehead to interpret emotions or moods.

Generally, the rapid growth and adoption of these techniques are attributed to the increasing availability of extensive facial databases, and improvements in facial recognition algorithms and their performance (Kaur et al., 2020; Tolba et al., 2005). The relevance of these technologies is evident in academic research, development and implementation, and in the business sector. For instance, several shopping centers in the United Kingdom have implemented digital signage systems capable of identifying customers' moods and ages, with the aim of delivering targeted advertising based on detected characteristics (Garaus et al., 2021; Nikolic, 2019), and the hospitality industry utilizes them to streamline the check-in process (Xu et al., 2021). Reflecting their significance, the facial recognition market value is projected to reach 19.3 billion US dollars by 2032, driven by advancements in AI (Borgeaud, 2024).

However, literature has indicated that some facial recognition systems lack the necessary accuracy and, in various cases, fail to correctly identify users (Zhou et al., 2023). This is partly due to the training of recognition algorithms on diverse image sets, leading to demographic biases stemming from statistical imbalances in the representation of certain groups within the datasets (Dominguez-Catena et al., 2024).

2.2. Facial Reading: Analyzing Consumer Emotions

Neuroscience has gained prominence due to the crucial role that emotions play in the human cognitive process (Hamelin et al., 2017). Growing evidence suggests that consumer emotions are predominantly influenced by implicit mechanisms, leading researchers to place increasing value on implicit measurements (Zhi et al., 2020). Specifically, the importance of the universality and uniqueness of facial expressions has been highlighted, positioning facial recognition as one of the leading biometric methods (Kaur et al., 2020). Contextually, facial reading techniques assist in analyzing and evaluating subjects' facial expressions under specific circumstances. Face reading is a somatic biometric method that uses the physical characteristics of the face to recognize physiological or behavioral aspects (Berčík et al., 2024). Among its advantages is that it is non-invasive, unlike techniques such as EEG.

Facial recognition systems follow a general processing flow; however, each software might present some differences. Kaur et al. (2020) identified five key steps: (1) image capture, (2) face detection, (3) feature extraction, (4) database matching, and (5) verification. Facial reading systems, in turn, include processes that allow for the extraction and analysis of individuals' facial expressions (Noussair et al., 2024; Shiau et al., 2023). Emotions can be detected through physical expressions and measured by physiological changes (Liu et al., 2021), and these can be influenced by various factors such as health, cultural expression, environment, etc. (Mumenthaler et al., 2020).

Different facial readers are based on Ekman's & Friesen's work (1976), which classifies six universal basic emotions: happiness, anger, sadness, fear, disgust, and surprise (Garaus et al., 2021). Various facial reading programs have been validated in the literature and applied in different contexts, like olfactory preferences (Berčík et al., 2024), public goods games (Noussair et al., 2024), outdoor experiences (Wei et al., 2021), and advertising (Hamelin et al., 2017).

Among the leading facial reading programs are Gfk-EMO Scan (Hamelin et al., 2017) and FireFACE (Wei et al., 2021), but the most widely used and validated across various studies is Noldus FaceReader. This software analyzes the basic emotions, and captures neutrality, arousal, and valence with 90% accuracy in classifying different expressions (Noussair et al., 2024). Its algorithm translates the vector of distances between over 500 facial points into a [0,1] scale, indicating how closely the facial expression aligns with profiles corresponding to the six basic universal emotions, neutrality, and arousal (Motro et al., 2021; Noussair et al., 2024). The program processes video recordings at 30 frames/second, allowing it to analyze each frame and produce a result where 0 indicates the absence of the emotion and 1 represents its maximum intensity (Noldus, 2024). Therefore, nonverbal methods like facial emotional expression analysis complement declarative approaches and yield valuable insights (Zhi et al., 2020). They provide a deeper understanding of emotions that might not be fully captured through self-reported data, allowing researchers to tap into implicit emotional responses. Consequently, integrating various techniques provides a more comprehensive assessment of consumer emotions and behavior.

Therefore, the versatility and usability of the face readers in marketing make it an especially valuable tool. Its ability to analyze consumers' emotions in real-time allows companies to gain deeper insights into audience reactions to various stimuli, including advertisements or products. This facilitates the creation of more refined and personalized strategies, enhances companies' ability to anticipate the needs and desires of their audience, leading to a more precise and effective approach to marketing decision-making.

3. Eye Tracking

3.1. What is Eye Tracking and Types of Eye Movements

Eye tracking is a method used to measure and analyze visual attention by recording eye movements and gaze patterns during specific tasks. It distinguishes between two types of attention control: top-down attention, driven by individual goals (endogenous attention), and bottom-up attention, which is stimulus-driven (exogenous attention) (Orquin & Loose, 2013). By monitoring the direction and location of eye movements, eye tracking offers insights into how attention is allocated, serving as a window into cognitive processes. The method has evolved significantly since its inception by Charles Bell in 1823, who linked eye movements to brain activity, laying the foundation for future studies connecting visual attention to neural and cognitive functions (Carter & Luke, 2020). This capability is particularly valuable in consumer research, as visual attention directly influences perception, information retention, and subsequent behavioral responses (Costa-Feito & Blanco-Moreno, 2023). Studies show that the areas where consumers focus most often—measured by fixations—tend to be those most relevant to their decision-making process, impacting product recall, preference formation, and purchase intention (Hannula et al., 2010; Muñoz-Leiva et al., 2018; Simonetti & Bigne, 2022). By observing these patterns, researchers gain an understanding of the cognitive pathways through which attention is transformed into consumer choices, allowing for the design of more effective visual stimuli that enhance consumer engagement and drive decisions.

The human eye has limited high-acuity vision, restricted to a small portion of the visual field known as the fovea, which necessitates frequent eye movements to capture relevant information. Eye tracking helps measure where and for how long individuals direct their gaze, offering valuable insights into cognitive processes, namely, attention, memory, and decision-making (Carter & Luke, 2020). The eye-mind link suggests that eye movements often mirror the ongoing mental activity, making this method a powerful tool in studies of attention and perception (Muñoz-Leiva et al., 2022). Modern eye-tracking techniques enable precise real-time analysis, improving the accuracy and accessibility of this tool for researchers and practitioners. Eye movements occur largely outside of conscious control, providing an unbiased, moment-by-moment reflection of cognitive engagement (Carter & Luke, 2020). They are categorized into two primary types: fixations and saccades.

- Fixations refer to moments when the eyes pause on a visual stimulus, allowing for information intake. They last between 180 to 330 milliseconds, depending on task complexity and individual differences (Rayner, 2009).
- Saccades are rapid, ballistic eye movements between fixations. During saccades, visual input, a phenomenon called saccadic suppression, is momentarily suppressed,.
- Other movements include smooth pursuit, used to follow moving objects, and vergence, which adjusts eye positioning when objects move closer or farther away.

3.2. How Eye Trackers Work and Eye-tracking Data

Modern eye trackers primarily rely on video-based systems which project invisible infrared light onto the eye, producing reflections from the cornea, which are recorded by specialized software. It also tracks the pupil, and by calibrating these data points, researchers can accurately estimate the point of gaze (Carter & Luke, 2020). Sampling rates for eye trackers vary widely, with high-end systems offering rates of up to 2000 Hz, capturing minute details of eye movement. In contrast, mobile and wearable eye trackers have lower sampling rates, suitable for real-world tasks but less precise in temporal accuracy (Andersson et al., 2010). Eye-tracking data typically consists of raw samples, each containing x and y coordinates of gaze positions. Depending on the sampling rate, hundreds or even thousands of data points are collected every second. This raw data is processed to identify key events like fixations, saccades, and blinks (Carter & Luke, 2020).

Researchers are interested in how often participants fixate on specific areas of a stimulus, known as regions or areas of interest (AOI). Through this process, eye tracking offers detailed data and strategic insights into how visual elements in advertisements, product displays, or web interfaces can influence attention and ultimately sway consumer choices (Alonso Dos Santos et al., 2019; Kong et al., 2019; Muñoz-Leiva et al., 2019). For instance, metrics including fixation duration and dwell time highlight which aspects of a visual display attract sustained consumer attention, reflecting deeper cognitive processing that often correlates with higher purchase likelihood or brand recall. Following Carter and

Luke (2020), common eye-tracking measures include fixation duration (length of time a participant focuses on a particular stimulus), saccade amplitude (distance traveled by the eye between fixations), time to first fixation (how long it takes for a participant to first fixate on a specific AOI), and dwell time (total time spent fixating within a given AOI). These can be further divided into early measures (time to first fixation and duration of first fixation), which assess initial attention to stimuli, and late measures (dwell time and number of fixations in an AOI), which provide insight into sustained attention and cognitive processing.

4. Electrodermal Activity

4.1. Conceptualization of Electrodermal Activity

Electrodermal activity is a response of the sympathetic system through sweating responses. Specifically, it translates into changes in the electrical characteristics of the skin in response to stimuli and during an alert situation (Białowąs & Szyszka, 2019; Kołodziej et al., 2019). Within the analysis of electrodermal activity, several metrics are considered and developed in behavioral studies, with electrodermal activity being the umbrella concept that encompasses the others. Electrodermal activity is a key indicator of emotional stimulation (Białowąs & Szyszka, 2019); it is a very useful metric for measuring emotional arousal as it has been shown to be effective in monitoring emotions and attention (Al-Nafjan et al., 2023; Casado-Aranda et al., 2023; Li et al., 2015; Lin et al., 2024). This relationship between sweating and emotionality is explained by the fact that brain activity in areas like the amygdala and the hippocampus are also related to eccrine control (Białowąs & Szyszka, 2019).

This metric has its advantages and disadvantages in the application to consumer behavioral studies. As advantages, it is a non-invasive, objective, and implicit measure, continuous and fast, and the cheapest and most appropriate methodologies to assess emotions implicitly and avoid biases (Caruelle et al., 2019; Casado-Aranda et al., 2023). This is why studies that apply electrodermal activity are numerous, making it a widely used method to assess the response of the ANS (Li et al., 2015; Walters et al., 2023). But for applying this methodology, it should be taken into account that it is not a dimensional emotional measure, as has been proven by authors like Kreibig (2010). It does not measure the valence of emotions, and cannot differentiate between positive or negative emotions (Caruelle et al., 2019; Li et al., 2015). Despite this characteristic, electrodermal response has been widely developed in marketing in retail, communication, design, consumer preference or shopping experience tracking (Białowąs & Szyszka, 2019). Recently, the application and use of these tools and metrics has been increasing due to the relevance of the information obtained (Białowąs & Szyszka, 2019; Caruelle et al., 2019).

Therefore, this tool is often complemented with other methodologies that qualify the information on the valence of these emotions, like electromyography (EMG), electroencephalography (EEG), or heart rate (HR) (Walters et al., 2023).

4.2. Development and Lines of Research

The study of electrodermal activity poses various methodological and application challenges. As for the metric itself, by nature, it is a rapid measurement that ensures the validity and reliability of data collection, implying that the response to different stimuli, at the dermal level, is translated into physiological reactions on the spot. This is an advantage over other metrics because it allows a more accurate measurement of these implicit reactions (Caruelle et al., 2019).

However, taking into account the characteristics of this metric, the basal data are essential in data collection, and must be collected in each condition and for each participant, being these the metrics that will allow the unification of the data of all the participants through emotional activation ratios (Li et al., 2022).

It is also a highly sensitive metric. This characteristic has two consequences: it ensures the reliability of data collection but it can also be affected or biased. Thus, the environment must be extensively controlled. In behavioral research, one of the lines of greatest interest is to measure implicit reactions in real situations, increasing the external validity of the studies (Li et al., 2022). In contrast, electrodermal activity is not a good metric for this type of study because control is lost. To improve this weakness, some companies are currently creating lighter hardware that ensures good data collection and is less intrusive to the participant. Among others, this is the case of Biopac's ResRing, which presents a novel, mobile, wearable device for measuring electrodermal activity, pulse, heart activity, and temperature (Biopac, 2024). These advances in measurement tools make it possible to extend the evaluation environments but always taking into account the restrictions and internal validity of the metric itself.

Regarding the new advances and incorporations in this type of biometric tools, the inclusion of new algorithms and technological improvements stands out. These include, for example, the option of taking data on mobile devices, making it easier to carry out studies in a real environment or to track them. Also, the simultaneous monitoring of different devices helps to comprehensively understand the emotional and attentional phenomenon. The supply of biometric data collection software takes into account this simultaneity by helping data collection and analysis of the specific characteristics of each metric.

For data analysis, statistical tools such as MatLab, through Signal Processing Toolbox, already include specific algorithms for the analysis of electrodermal data, which facilitates the cleaning, filtering, and management of these data, which generally refer to a large volume of data (MatLab, 2024). Additionally, new ML and deep learning applications are being used to train computers to distinguish patterns in a similar way to the human brain, improving the understanding and study of these naturally complex consumer decision-making and behavioral processes (Rodgers et al., 2021).

Among the technological advances, the inclusion of immersive tools such as virtual reality, augmented reality, or mixed reality can improve the external validity of behavioral experiments, while maintaining control over space and external variables (Al-Nafjan et al., 2023).

5. New Horizons and Limitations

Athough these tools have advanced significantly, they still face important limitations that must be addressed to reach their full potential. Demographic bias remains a critical challenge, affecting the accuracy of these systems for certain population groups (Dominguez-Catena et al., 2024).

In terms of facial reader tools, the reliability of results can be compromised by external factors like image quality or lighting conditions, which limits their applicability in real-world scenarios (Kaur et al., 2020; Shiau et al., 2023). Concerns about the privacy and security of data collected through facial recognition raise serious ethical dilemmas (Gies et al., 2020), which are further intensified by the lack of a clear and uniform regulatory framework.

For electrodermal activity, the sensitivity of the tool itself may be a limitation. The need to control all independent and external conditions and variables to ensure correct data collection leads to fictitious scenarios of low external validity. Additionally, the lack of consensus on how to treat this metric statistically means that the results of the studies may be biased. The systematization of the protocols and the sophistication of the research allows improving the experimental and behavioral analysis processes with these tools (Caruelle et al., 2019).

Muñoz-Leiva et al. (2022) argue that eye tracking has been applied across diverse fields, namely, psychology, medicine, business, engineering, education, environmental studies, and human-computer interaction. Its impact has been particularly significant in marketing research, where it is frequently utilized within consumer neuroscience to analyze how advertising and other stimuli influence consumer behavior. The authors emphasize that future research is likely to advance our comprehension of the relationships between attention, memory, and emotions, and how these factors interact with consumer beliefs and decision-making processes. Furthermore, there is an increasing interest in integrating eye tracking with other physiological measures, which promises to offer a more comprehensive perspective on cognitive and emotional responses. However, challenges remain, including the need for greater precision in interpreting data, ethical considerations regarding privacy, and the complexity of isolating eye movements from other cognitive processes.

These challenges open new horizons for research and development. The future of these technologies will depend on the ability to develop more inclusive algorithms, improve accuracy under diverse conditions, and establish regulations that protect individual privacy. With these advancements, biometric technologies could become highly effective tools. Additionally, biometric tools have the potential to become a key tool in marketing studies, enabling companies to gain a deeper and more nuanced understanding of consumer emotional and attentional responses, which could transform the way marketing strategies are conceptualized and optimized.

Acknowledgments

This work has been part of the project TED2021-129513B-C21, funded by MCIN/AEI/10.13039/501100011033 and by the European Union "NextGenerationEU"/PRTR.

References

Al-Nafjan, A., Aldayel, M., & Kharrat, A. (2023). Systematic review and future direction of neuro-tourism research. *Brain Sciences*, 13(4), 682. https://doi.org/10.3390/brainsci13040682.

Alonso Dos Santos, M., Calabuig Moreno, F. & Crespo-Hervás, J. (2019). Influence of perceived and effective congruence on recall and purchase intention in sponsored printed sports advertising: An eye-tracking application. *International Journal of Sports Marketing and Sponsorship*, 20(4), 617–633. https://doi.org/10.1108/IJSMS-10-2018-0099.

Andersson, R., Nystrom, M., & Holmqvist, K. (2010). Sampling frequency and eye-tracking measures: How speed affects durations, latencies, and more. *Journal of Eye Movement Research*, 3(3). https://doi.org/10.16910/jemr.3.3.6.

Berčík, J., Mravcová, A., Sendra Nadal, E., López Lluch, D.B., & Farkaš, A. (2024). FaceReader as a neuromarketing tool to compare the olfactory preferences of customers in selected markets. *Spanish Journal of Marketing – ESIC* (ahead-of-print). https://doi.org/10.1108/SJME-03-2023-0050.

Białowąs, S., & Szyszka, A. (2019). Measurement of electrodermal activity in marketing research. In: *Managing Economic Innovations: Methods and Instruments*, 73–90. ResearchGate. https://doi.org/10.12657/9788379862771-5.

Biopac (2024, August 30). *Research Ring*. BIOPAC Systems Inc. https://www.biopac.com/product/research-ring/

Borgeaud, A. (2024, July 1). *Facial Recognition Global Market Size 2032*. Statista. https://www.statista.com/statistics/1153970/worldwide-facial-recognition-revenue/

Carter, B. & Luke, S. (2020). Best practices in eye tracking research. *International Journal of Psychophysiology*, 155, 49–62. https://doi.org/10.1016/j.ijpsycho.2020.05.010.

Caruelle, D., Gustafsson, A., Shams, P., & Lervik-Olsen, L. (2019). The use of electrodermal activity (EDA) measurement to understand consumer emotions: A literature review and a call for action. *Journal of Business Research*, 104, 146–160. https://doi.org/10.1016/j.jbusres.2019.06.041.

Casado-Aranda, L.A., & Sánchez-Fernández, J. (2022). Advances in neuroscience and marketing: Analyzing tool possibilities and research opportunities. *Spanish Journal of Marketing – ESIC*, 26(1), 3–22. https://doi.org/10.1108/SJME-10-2021-0196.

Casado-Aranda, L.A., Sánchez-Fernández, J., Bigne, E., & Smidts, A. (2023). The application of neuromarketing tools in communication research: A comprehensive review of trends. *Psychology & Marketing*, 40(9), 1737–1756. https://doi.org/10.1002/mar.21832.

Costa-Feito, A., & Blanco-Moreno, S. (2023). Unconscious and conscious aspects of healthy food consumption: A neuromarketing and artificial intelligence approach. In: Benítez-Andrades, J.A., García-Llamas, P., Taboada, Á., Estévez-Mauriz, L., & Baelo, R. (Eds.), *Global Challenges for a Sustainable Society.* EURECA-PRO 2022. Springer Proceedings in Earth and Environmental Sciences. Springer, Cham. https://doi.org/10.1007/978-3-031-25840-4_35.

Dominguez-Catena, I., Paternain, D., & Galar, M. (2024). Metrics for dataset demographic bias: a case study on facial expression recognition. *IEEE Transactions on Pattern Analysis and Machine Intelligence*, 46(8), 5209–5226. https://doi.org/10.1109/TPAMI.2024.3361979.

Ekman, P., & Friesen, W.V. (1976). Measuring facial movement. *Environmental Psychology & Nonverbal Behavior*, 1(1), 56–75. https://doi.org/10.1007/BF01115465.

Garaus, M., Wagner, U., & Rainer, R.C. (2021). Emotional targeting using digital signage systems and facial recognition at the point-of-sale. *Journal of Business Research*, 131, 747–762. https://doi.org/10.1016/j.jbusres.2020.10.065.

Gies, W., Overby, J., Saraceno, N., Frome, J., York, E., & Salman, A. (2020). Restricting data sharing and collection of facial recognition data by the consent of the user: A systems analysis. *2020 Systems and Information Engineering Design Symposium* (*SIEDS*), 1–6. https://doi.org/10.1109/SIEDS49339.2020.9106661.

Hamelin, N., Moujahid, O.E., & Thaichon, P. (2017). Emotion and advertising effectiveness: A novel facial expression analysis approach. *Journal of Retailing and Consumer Services*, 36, 103–111. https://doi.org/10.1016/j.jretconser.2017.01.001.

Hannula, D.E., Althoff, R.R., Warren, D.E., Riggs, L., Cohen, N.J., & Ryan, J.D. (2010). Worth a glance: Using eye movements to investigate the cognitive neuroscience of memory. *Frontiers in Human Neuroscience*, 4(166). https://doi.org/10.3389/fnhum.2010.00166.

Kaur, P., Krishan, K., Sharma, S.K., & Kanchan, T. (2020). Facial-recognition algorithms: A literature review. *Medicine, Science and the Law*, 60(2), 131–139. https://doi.org/10.1177/0025802419893168.

Kong, S., Huang, Z., Scott, N., Zhang, Z., & Shen, Z. (2019). Web advertisement effectiveness evaluation: Attention and memory. *Journal of Vacation Marketing*, 25(1), 130-146. https://doi.org/10.1177/1356766718757272

Kreibig, S.D. (2010). Autonomic nervous system activity in emotion: A review. *Biological Psychology*, 84(3), 394–421. https://doi.org/10.1016/j.biopsycho.2010.03.010.

Lajante, M., & Lux, G. (2020). Perspective: Why organizational researchers should consider psychophysiology when investigating emotion? *Frontiers in Psychology*, 11, 1–7. https://doi.org/10.3389/fpsyg.2020.01705.

Levenson, R.W. (2014). The autonomic nervous system and emotion. *Emotion Review*, 6(2), 100–112. https://doi.org/10.1177/1754073913512003.

Li, S. (2019). Emotional appeals in tourism TV commercials: A psycho-physiological study. *Journal of Hospitality and Tourism Research*, 43(6), 783–806. https://doi.org/10.1177/1096348019828440.

Li, S., Scott, N., & Walters, G. (2015). Current and potential methods for measuring emotion in tourism experiences: A review. *Current Issues in Tourism*, 18(9), 805–827. https://doi.org/10.1080/13683500.2014.975679.

Li, S., Sung, B., Lin, Y., & Mitas, O. (2022). Electrodermal activity measure: A methodological review. *Annals of Tourism Research*, 96, 103460. https://doi.org/10.1016/j.annals.2022.103460.

Lin, Y., Mitas, O., Shen, Y., Bastiaansen, M., & Strijbosch, W. (2024). Objective measurement of experiences in tourism and hospitality: A systematic review of methodological approaches and best practices. *Journal of Hospitality and Tourism Research*. https://doi.org/10.1177/10963480231226086.

Liu, P., Liu, M., Xia, T., Wang, Y., & Guo, P. (2021). The relationship between landscape metrics and facial expressions in 18 urban forest parks of northern China. *Forests*, 12(12), Article 12. https://doi.org/10.3390/f12121619.

MatLab (2024, August 30). *Signal Processing Toolbox*. MathWorks. https://es.mathworks.com/products/signal.html.

Motro, D., Comer, D.R., & Lenaghan, J.A. (2021). Examining the effects of negative performance feedback: The roles of sadness, feedback self-efficacy, and grit. *Journal of Business and Psychology*, 36(3), 367–382. https://doi.org/10.1007/s10869-020-09689-1.

Mumenthaler, C., Sander, D., & Manstead, A.S.R. (2020). Emotion recognition in simulated social interactions. *IEEE Transactions on Affective Computing*, 11(2), 308–312. https://doi.org/10.1109/TAFFC.2018.2799593.

Muñoz-Leiva, F., Liébana-Cabanillas, F. & Hernández-Méndez, J. (2018). E-tourism advertising effectiveness: Banner type and engagement as moderators. *Journal of Services Marketing*, 32(4), 462–475. https://doi.org/10.1108/JSM-01-2017-0039.

Muñoz-Leiva, F., Hernández-Méndez, J. & Gómez-Carmona, D. (2019). Measuring advertising effectiveness in Travel 2.0 websites through eye-tracking technology. *Physiology & Behavior*, 200, 83–95. https://doi.org/10.1016/j.physbeh.2018.03.002.

Muñoz-Leiva, F., Rodríguez-López, M.E. & García-Martí, B. (2022). Discovering prominent themes of the application of eye tracking technology in marketing research. *Management Letters*, 22(1), 97–113. https://doi.org/10.5295/cdg.211516fm.

Nikolic, I. (2019). Billboards spying on shoppers can detect their ages and mood. *Dailymail*. https://www.dailymail.co.uk/news/article-6969213/Billboards-spying-shoppers-facial-recognition-cameras-detect-ages-mood.html.

Noldus. (2024). *Facial Expression Recognition Software | FaceReader* [Computer software]. Noldus. https://www.noldus.com/facereader.

Noussair, C.N., Tucker, S., Xu, Y., & Breaban, A. (2024). The role of emotions in public goods games with and without punishment opportunities.

Journal of Economic Behavior & Organization, 217, 631–646. https://doi.org/10.1016/j.jebo.2023.11.003.

Orquin, J. & Loose, S. (2013). Attention and choice: A review on eye movements in decision making. *Acta Psychologica*, 144(1), 190–206. https://doi.org/10.1016/j.actpsy.2013.06.003.

Rayner, K. (2009). Eye movements and attention in reading, scene perception, and visual search. *Quarterly Journal of Experimental Psychology*, 62(8), 1457–1506.

Rodgers, W., Yeung, F., Odindo, C., & Degbey, W.Y. (2021). Artificial intelligence-driven music biometrics influencing customers' retail buying behavior. *Journal of Business Research*, 126, 401–414. https://doi.org/10.1016/j.jbusres.2020.12.039.

Shiau, W.-L., Liu, C., Zhou, M., & Yuan, Y. (2023). Insights into customers' psychological mechanism in facial recognition payment in offline contactless services: Integrating belief–attitude–intention and TOE–I frameworks. *Internet Research*, 33(1), 344–387. https://doi.org/10.1108/INTR-08-2021-0629.

Simonetti, A., & Bigne, E. (2022). How visual attention to social media cues impacts visit intention and liking expectation for restaurants. *International Journal of Contemporary Hospitality Management*, 34(6), 2049–2070. https://doi.org/10.1108/IJCHM-09-2021-1091.

Tolba, A., El-Baz, A., & El-Harby, A. (2005). Face recognition: A literature review. *International Journal of Signal Processing*, 2, 88–103.

Vila, J. (1996). *Una introducción a la psicofisiología clínica.* Ediciones Pirámide.

Walters, G., Jiang, Y., & Li, S. (2023). Physiological measurements in hospitality and tourism research: A systematic review and new theoretical directions. *Journal of Hospitality & Tourism Research.* https://doi.org/10.1177/10963480231199990.

Wei, H., Hauer, R.J., & He, X. (2021). A forest experience does not always evoke positive emotion: A pilot study on unconscious facial expressions using face reading technology. *Forest Policy and Economics*, 123, 102365. https://doi.org/10.1016/j.forpol.2020.102365.

Xu, F.Z., Zhang, Y., Zhang, T., & Wang, J. (2021). Facial recognition check-in services at hotels. *Journal of Hospitality Marketing & Management*, 30(3), 373–393. https://doi.org/10.1080/19368623.2020.1813670.

Zhi, R., Hu, X., Wang, C., & Liu, S. (2020). Development of a direct mapping model between hedonic rating and facial responses by dynamic facial expression representation. *Food Research International*, 137, 109411. https://doi.org/10.1016/j.foodres.2020.109411.

Zhou, Y., Liang, Y., & Tan, P. (2023). Design of an intelligent laboratory facial recognition system based on expression keypoint extraction. *IEEE Access*, 11, 129805–129817. https://doi.org/10.1109/ACCESS.2023.3329575.

4

Biometric Authentication Systems in Banking: Improving Security and Customer Experience while Fighting Cybercrime

Francisco Liébana-Cabanillas[1*(0000-0002-3255-0651)], Juan F. Prados-Castillo[1(0000-0003-3588-5534)], Elena Higueras-Castillo[1(0000-0002-2130-5212)] and Zoran Kalinic[2(0000-0001-8137-9005)]

[1] University of Granada
[2] University of Kragujevac

1. Introduction

The increasing digitization of financial services, accompanied by the proliferation of increasingly sophisticated cyberattacks, poses an urgent need to strengthen security systems in the banking sector. Conventional authentication methods, including passwords and PINs, are vulnerable to attacks like phishing and spoofing and are insufficient to ensure security in a constantly evolving digital environment (Venkatraman & Deltahedra, 2008). Contextually, biometric authentication is an innovative solution with the potential to substantially improve security and simultaneously optimize customer experience. By relying on the unique and non-transferable characteristics of individuals, namely, fingerprints and facial features, biometrics increase protection against fraud and facilitate faster and more convenient authentication processes, eliminating the need to remember passwords or access codes (Jain & Ross, 2016).

*Corresponding author: franlieb@ugr.es

Biometric authentication systems are increasingly being adopted in banking to improve security and user convenience. They use unique physical or behavioral characteristics, namely, fingerprints, facial features, voice, and iris patterns for identification (Asl et al., 2021; Dobbie, 2020). Biometrics offer greater security than traditional methods, like PINs and passwords, which are vulnerable to fraud and breaches (Iyawe, 2010; Venkatraman & Delpachitra, 2008). Multiple biometric factors can be combined to improve online banking security (Venugopal & Viswanath, 2016). Biometric authentication helps combat cybercrime and terrorism by preventing identity theft and unauthorized access (Thakur et al., 2024). However, the implementation of biometric systems presents challenges, namely, technological, managerial, social, and ethical considerations (Venkatraman & Delpachitra, 2008). Despite these challenges, biometrics are seen as the future of user authentication in financial services, offering greater security, faster transactions, and better customer experience (Fatima, 2011).

This chapter provides a comprehensive and up-to-date perspective on the impact of biometrics in banking to understand the specific challenges associated with its implementation, namely, the costs and privacy concerns of biometric data (European Union, 2016). It also seeks to provide a theoretical and practical framework to enable financial institutions to assess the benefits and limitations of this technology and serve as a basis for future research and development in the field of authentication in banking. The main objective is to examine the impact of biometric authentication systems in the banking sector, focusing on their role in enhancing security and enriching customer experience (Ross et al., 2008). The specific objectives of this study are:

RO1: We evaluate how biometric technologies, compared to traditional authentication methods, contribute to reducing fraud and cyberattacks within the banking sector. This evaluation aims to provide empirical evidence on the effectiveness of technologies like fingerprint and facial recognition in countering threats including phishing and spoofing (Venkatraman & Delpachitra, 2008; Basare et al., 2023).

RO2: We evaluate the benefits to customer experience by examining the impact of biometric authentication on user experience, specifically speed, convenience, and perceived security. How these technologies can improve customer satisfaction through simplified and efficient authentication are analyzed (Jain, 2007; Lucia et al., 2023).

RO3: The main barriers to the adoption of biometric systems in financial institutions are researched, including analysis of technological barriers, implementation costs, and concerns regarding the privacy of biometric data to provide a comprehensive understanding of the factors limiting their adoption (Dharavath et al., 2013; Ayeswarya & Singh, 2024).

RO4: Case studies and best practices of financial institutions that have successfully implemented biometric authentication, explored effective practices, and analyzed challenges overcome are presented. They provide a framework for institutions interested in adopting these technologies (Dobbie, 2020).

RO5: The future potential of biometrics in the banking sector is analyzed by examining the emerging role of biometrics as a possible authentication standard in banking. As technological advances and regulatory demands call for greater security, the potential for biometrics to become the norm within the financial industry are examined (Marani et al., 2023).

2. Theoretical Framework

2.1. Definition of Biometric Authentication Systems

Biometric authentication systems are technologies that make it possible to verify an individual's identity through unique and non-transferable physiological or behavioral characteristics (Basare et al., 2023; Bhartiya et al., 2018). Unlike traditional methods, like passwords or PINs, which can be forgotten, shared, or stolen, biometrics use personal traits that are difficult to replicate, which significantly increases the security of authentication processes (Jain, 2007). Biometric authentication is based on the recognition of physical characteristics, namely, fingerprints, iris patterns, facial features, vein patterns, and behavioral characteristics including voice, gait, and handwriting rhythm (Tripathi & Nishad, 2020). These technologies process and store biometric information in secure databases and compare it each time a user attempts to authenticate it in a system. Because it is extremely difficult to forge, biometric information provides a higher level of security than traditional authentication methods (Kataria et al., 2013).

Biometric systems are classified into physiological and behavioral biometrics. Physiological biometrics focus on physical traits of the body, namely, fingerprints, face, and iris (Kataria et al., 2013; Tripathi & Nishad, 2020), while behavioral biometrics focuses on specific behavioral patterns of the individual, like typing dynamics or gait. Each method has its own applications, advantages, and challenges, and can be employed independently or in combination to increase security in critical systems, especially in the financial sector (Bhartiya et al., 2018). The implementation of biometric authentication in financial systems aims to improve security and offers a more convenient and efficient user experience. By reducing reliance on traditional methods, biometrics offer fast authentication, which is especially important in mobile applications and digital transactions. The integration of artificial intelligence (AI) and advanced recognition algorithms has improved the accuracy and speed of these systems, making them increasingly accessible and efficient (Asl et al., 2021).

Currently, researchers are exploring multimodal systems, continuous authentication, and AI integration to improve accuracy and security (Tripathi & Nishad, 2020). Biometric authentication finds applications in access control, financial transactions, healthcare, and law enforcement. As the field evolves, addressing ethical and legal issues, improving privacy protection, and integration with emerging technologies like IoT remain crucial areas for future research (Dharavath et al., 2013).

2.2. Types of Biometric Technologies Applied in Banks

Biometric technologies are increasingly being adopted in the banking sector to improve security and user convenience. They have transformed the way financial institutions authenticate the identity of their customers, thus improving the security and efficiency of transactions. Several authors have proposed a classification of these technologies (Barkadehi et al., 2018; Smith-Creasey et al., 2023; Ryu et al., 2023; Liébana-Cabanillas et al., 2024), but we will identify facial recognition, fingerprint, vein, and iris recognition, voice authentication, retina and DNA identification, and behavioral biometrics (Figure 1).

Facial Recognition

Face recognition involves capturing and analyzing the distinctive features of a person's face (Nakisa et al., 2023). This process is generally performed by comparing an image or video of a user with a database of previously recorded faces. In banking, this technology is used to verify the identity of customers during the online account opening process and authorize financial transactions. The implementation of advanced

Figure 1: Biometric technologies. *Source:* Own elaboration from Pixabay.

AI algorithms enables accurate identification, even under varying lighting conditions or changes in the user's appearance (Khare et al., 2024). Facial recognition also facilitates a frictionless user experience by eliminating the need to remember passwords and additional devices for authentication (Sudhakar et al., 2024).

Fingerprint Recognition

Fingerprint recognition is one of the oldest and most widely adopted methods for biometric authentication. It is based on capturing the user's fingerprint and comparing it with stored records. In the banking context, it is used in ATMs and mobile applications to authenticate customers before they perform transactions (Ali et al., 2023). The uniqueness of an individual's fingerprint provides a high level of security. However, factors such as finger injuries or skin conditions can affect recognition accuracy (Sumalatha et al., 2024).

Voice Authentication

Voice authentication analyzes unique characteristics of speech, including pitch, frequency, and rhythm, to verify the user's identity. It is particularly useful in telephone-banking services and mobile applications that allow voice commands. Voice biometrics offer a convenient user experience, as verification occurs naturally during conversation (Singh et al., 2018). However, factors such as environmental noise, voice-affecting diseases, and impersonation attempts using recordings can influence the effectiveness of this method.

Iris Recognition

Iris recognition is based on the identification of unique patterns in the colored part of the human eye. This method is highly accurate owing to the complexity and stability of iris patterns over time (Rashidi et al., 2024; Al-Debei et al., 2024). In the banking sector, it is used in applications that require a high level of security, including access to safety deposit boxes or authorization of high-value transactions. Although offering superior security, it is costly to implement and requires specialized equipment, limiting its widespread adoption.

Retinal Identification

Retinal identification is based on the analysis of unique patterns of blood vessels at the back of the eye. It offers an extremely high level of security owing to the complexity and uniqueness of these patterns. In banking, this is used in environments that require maximum security, namely, access to highly sensitive financial information (Mubeen, 2024). However, its implementation is limited owing to the high costs and specialized equipment needed to perform the scans.

Vein Recognition

Vein recognition involves capturing unique patterns from the veins of the palm or fingers using infrared technology (Mubeen, 2024). It is difficult to spoof because the vein patterns are located beneath the skin and are externally invisible. In banking, it has been implemented in ATMs and access control systems to authenticate users (Boudjellal, 2024). The system's accuracy and security makes it suitable for critical applications, although its adoption may be limited by costs and specialized hardware.

DNA Identification

DNA identification is the most accurate biometric method and is based on the comparison of an individual's unique genetic material (Hassan & Sabahat, 2024; Mohammed et al., 2024). Although common in forensic applications, its use in banking could be relevant in the future to ensure maximum security in critical transactions. This method has significant limitations in terms of time, cost, and technical complexity, hindering its widespread adoption in the financial sector.

Behavioral Biometrics

Behavioral biometrics analyzes user behavior patterns, like keyboard typing, mouse movement, or interaction with mobile devices. It is used to detect suspicious activities and prevent fraud, complementing other authentication systems (Elhag et al., 2024). In the banking sector, it allows real-time monitoring and analysis of user interactions, identifying anomalies that could indicate fraud attempts (Wang et al., 2024). Its implementation does not require additional hardware and can be integrated into existing digital platforms, offering an additional layer of security without affecting the user experience.

The adoption of biometric authentication systems in the banking sector improves the security of financial transactions and optimizes the customer experience by facilitating access to services and reducing reliance on traditional authentication methods. It is critical to address privacy and data protection considerations to ensure that the implementation of these technologies complies with current regulations and respects user rights.

2.3. Importance of Biometric Authentication for Bank Security

Biometric authentication has become a key pillar for ensuring security in the banking sector, especially in environments where cyber threats and financial fraud are constantly increasing. This technological approach provides identity verification methods based on unique physical and behavioral characteristics, which significantly reduce the risk of unauthorized access and other malicious activities (Khan et al., 2023). Along these lines, several elements confirm the importance of the relationship between biometric authentication and banking security (Choquehuanca-Chuctaya & Arroyo-Paz, 2024).

A major benefit of biometric authentication is its ability to prevent fraud. Traditional authentication methods, namely, passwords and PIN codes, are vulnerable to attacks like phishing or social engineering. In contrast, biometric characteristics, namely, fingerprints, facial recognition, or vein patterns are unique and intrinsic to each individual, making them virtually impossible to replicate (Normalini & Ramayah, 2012). Facial recognition has proven to be particularly effective in detecting phishing attempts in mobile banking applications. The implementation of biometric technologies improves security and increases customer confidence in banking services. Recent studies have shown that users perceive biometric systems as more secure and convenient than traditional authentication methods (Piotrowska, 2024). This level of trust is crucial to strengthen the customer-bank relationship, especially in the digital age, where online transactions dominate the financial landscape.

Although the initial implementation of biometric technology may require significant investment, the long-term benefits of reduced operational costs are considerable. By reducing fraud and security-related disputes, banks can save significant resources that would otherwise be spent on loss recovery or litigation (Yang et al., 2021). Additionally, biometrics simplify the authentication processes, reducing the time and costs associated with customer service. Thus, increasing regulatory pressure in the financial arena has driven the adoption of biometric technologies. Regulations like the General Data Protection Regulation (GDPR) in Europe or the Electronic Funds Transfer Act (EFTA) in the United States emphasize the need for robust and secure authentication systems. Biometric technologies meet these requirements by offering advanced levels of data protection and transaction security (Poullet, 2023).

Despite its benefits, biometric authentication poses challenges related to privacy and data storage. The collection and storage of biometric information requires highly secure systems to prevent potential data leakage. It is also essential to ensure that the use of this technology respects the rights of users and complies with data protection laws like the GDPR (Cross et al., 2023).

3. Benefits of Biometric Systems in Financial Institutions

3.1. Improved Security: Reduction of Cybercrime (Phishing, Spoofing, etc.)

The incorporation of biometric systems into the financial sector has established itself as one of the most effective solutions for strengthening security against cybercrime. These systems act as an advanced barrier against threats, namely, phishing, spoofing, etc., that exploit vulnerabilities in traditional authentication methods. Unlike physical passwords or tokens, which are susceptible to being stolen, guessed, or shared, biometric characteristics are inherent to the user. These attributes make them difficult to replicate or forge, and eliminate reliance on stored credentials, reducing the risk of exposure to security breaches. Technologies such as liveness detection further strengthen these systems by verifying that biometric signals originate from a real user and not from simulations or recordings. This approach prevents spoofing attempts, ensuring that legitimate cardholders (Jain et al., 2016) only make transactions.

The impact of biometrics in mitigating cybercrime is supported by relevant data. According to IBM Security (2020), 20% of security breaches in the financial sector are attributed to compromised credentials. This recurring problem can be drastically reduced through the adoption of biometric systems, as personal characteristics cannot be stolen or reused by malicious actors. These systems also enable continuous monitoring and real-time detection of suspicious activities, enhancing the ability of institutions to proactively respond to emerging threats.

The implementation of biometrics addresses security challenges and helps build customer confidence by ensuring that their assets and personal data are protected. Financial institutions that have adopted these technologies reported a significant decrease in fraud incidents, consolidating their positions as leaders in innovation and financial security (National Security Institute, 2024).

3.2. Improved Customer Experience: Speed and Convenience

Biometric authentication has revolutionized customer interaction with financial services by providing a faster, more convenient, and efficient experience. Using biometrics, customers can access their accounts or complete transactions immediately, thereby simplifying the process and saving time.

Due to their ease of use, biometric systems integrate seamlessly with digital platforms, including mobile applications and advanced ATMs. For example, facial recognition allows customers to log in simply by looking at their device, whereas fingerprint scanning provides fast, one-touch authentication. This convenience significantly improves customer perception of the financial institution, positioning it as innovative and focusing on user needs.

The positive impact of biometrics on customer experience is supported by data. According to a Deloitte report (2021), 79% of consumers prefer to use biometric authentication to access mobile banking applications, highlighting security and convenience as the main reasons. This preference facilitates the recurrent use of these platforms and strengthens the relationship between the customer and the institution, fostering loyalty and satisfaction (Piotrowska et al., 2017).

Biometrics also removes accessibility barriers for various segments of the population, including those with difficulty in remembering passwords or handling additional devices such as security tokens. By ensuring a seamless and hassle-free experience, financial institutions enhance their value proposition and o position themselves to attract and retain a more diverse customer base (Alrawili et al., 2024).

In a competitive environment where service-based differentiation is the key, adoption of biometric technologies represents a significant strategic advantage. By offering a combination of speed, ease, and security (Kuznetsov, et al., 2024), these solutions transform the customer experience, and strengthen the position of financial institutions as leaders in innovation.

3.3. Reduced Fraud and Increased Customer Confidence

The incorporation of biometric systems in banking has proven to be a key tool for significantly reducing fraud cases while simultaneously strengthening customer confidence in financial institutions. By using unique and non-transferable user characteristics, namely, fingerprints, iris patterns, or facial recognition, these systems eliminate the vulnerabilities associated with traditional authentication methods, like passwords or PIN codes, which can be stolen or guessed.

A study by Javelin Strategy & Research (2020) revealed that financial institutions that adopted biometric authentication experienced a 50% reduction in identity fraud cases. These data highlight the direct impact of biometrics in protecting customers and financial assets. By ensuring that only legitimate users can access their accounts, these technologies minimize the risk of fraudulent transactions and reinforce the customers' perception of security.

Additionally, biometric systems play a key role in complying with strict regulations, such as know your customer (KYC) and anti-money laundering (AML) policies. These regulations require robust identity-verification processes to prevent illicit activities and ensure transparency in financial transactions. Biometrics facilitates this compliance by providing accurate and automated verification of customer identity, improving operational efficiency, and reducing the risk of regulatory penalties (Aratek, 2023).

Customer trust is a critical element in the financial sector, and the implementation of biometric systems contributes significantly to its strengthening. Knowing that their accounts are protected by advanced technologies, customers develop greater loyalty to institutions that prioritize their security, thus, creating a virtuous cycle in which fraud reduction and positive customer perception translate into a stronger and longer-lasting relationship.

In a context where cyber threats continue to evolve, biometrics is a solution to today's challenges and a strategic investment to ensure the security and trust of the financial system in the future.

4. Challenges and Limitations of Biometrics Implementation in Banks

4.1. Technological Barriers

Despite significant advances in biometric technology, its implementation in the financial sector faces several significant technical challenges (Ryu et al., 2023). A prominent problem is the accuracy of the system under varying conditions. Technologies such as facial recognition, which are effective in controlled environments, can present limitations in environments with poor lighting, unusual capture angles, or when users experience changes in their physical appearance (Romero, 2018). These inconsistencies can generate false-positive or false-negative results, affecting the confidence and functionality of the system.

Another challenge is the initial configuration of biometric systems. This process, which requires the capture of unique biometric data from each user to create an authentication pattern, can be laborious and consumes significant resources. The quality of the data collected at this stage is crucial because it determines the accuracy and reliability of the system over time. It is particularly challenging in institutions with a large customer base because it slows technology adoption.

Another critical aspect is the interoperability between platforms and systems. The diversity of vendors and technologies used by financial institutions makes it difficult to integrate biometric data into a unified environment. The lack of common standards complicates the creation of an interoperable platform where biometric data can be efficiently integrated and used alongside other datasets. This challenge is intensified in the case of national biometric identification projects, which require systems capable of operating on a large scale and adapting to different technological infrastructures (Soffid, 2023).

Additionally, ensuring the security and privacy of biometric data within integrated systems is a complex task. Banks must implement advanced encryption technologies and strict access controls to protect biometric information from potential security breaches.

Despite these barriers, technological advances continue to reduce these limitations. Improvements in AI algorithms, more sophisticated hardware, and the increasing adoption of international standards are helping overcome these challenges (Awadallah et al., 2024). However, for biometrics to reach its full potential in the financial sector, it is crucial to address these limitations proactively, ensuring the accuracy, interoperability, and security of the systems.

4.2. Privacy Challenges and Biometric Data Management

The implementation of biometric systems in the financial sector faces significant challenges related to the privacy and management of sensitive data. Biometric data, being unique and unalterable, represent a valuable but also extremely sensitive asset. Their storage, transfer, and use raise legal, ethical, and technical concerns that complicate the widespread adoption of these technologies.

One main challenge is ensuring security in the handling of biometric data. Financial institutions must implement advanced measures to protect this information against possible security breaches or misuses. This includes the use of robust encryption techniques, strict access controls, and regular audits to detect vulnerabilities. However, even with these measures, cyberattacks targeting biometric databases pose a significant threat (Soffid, 2023).

The collection of biometric data also raises ethical concerns. Users must be clearly informed about how their data are collected, stored, and used, and they must be able to grant or withdraw their consent at any time. The lack of transparency in this process can erode consumer confidence and generate resistance to the use of these technologies. These concerns were identified. A study by Ponemon Institute (2019) revealed that 69% of consumers expressed concerns about the privacy of their biometric data when used for financial authentication suggesting that certain segments of the population may be reluctant to adopt biometric technologies, particularly if they do not perceive clear guarantees of data protection.

Additionally, privacy regulations, such as the GDPR in Europe, impose strict requirements on the handling of biometric data. Financial institutions must comply with regulations that require transparency, data minimization, and limited use of collected information. This adds an additional layer of complexity to the implementation of these technologies, particularly in international environments with different regulatory frameworks.

To overcome these challenges, institutions must combine advanced technologies with ethical data management strategies, including educating users about the benefits and security measures associated with biometrics, ensuring regulatory compliance, and developing systems that prioritize privacy. This approach helps address existing concerns and builds user trust and acceptance.

4.3. Implementation and Adaptation Costs

The adoption of biometric systems in financial institutions represents a considerable investment, encompassing multiple areas beyond the acquisition of technology. Initial costs include the purchase of advanced hardware, like facial recognition cameras, fingerprint scanners, or iris readers, and the implementation of specialized software to enable the integration of these technologies into existing banking systems. However, these costs are only a part of the challenge.

One of the biggest challenges is the need to adapt existing technological infrastructure to support new biometric systems. This may involve upgrading networks, servers, and databases to handle the secure processing and storage of large volumes of biometric data. Staff training is essential to ensure that employees can operate and maintain these systems efficiently (Kantaros, 2023). This includes technical training and training on privacy issues and ethical data management.

Ongoing technical support is another factor that should be considered. Biometric systems require regular maintenance to ensure accuracy and reliability over time, including updating the software to prevent vulnerabilities and replacing the worn hardware, which adds a long-term recurring cost.

A report by Goode Intelligence (2020) projects that the global market for biometrics in financial services will reach $15 billion by 2025, reflecting the magnitude of the investments needed to implement these technologies on a large scale. However, the return on this investment may take time to materialize, especially in smaller institutions or in regions with limited resources. For these entities, initial costs can be a significant barrier, limiting their ability to compete with larger institutions that have greater financial resources.

Despite these costs, potential benefits, including reduced fraud, improved security, and enhanced customer experience, may justify investment in the long term. To mitigate financial challenges, many institutions choose to implement biometric solutions incrementally, starting with key areas like mobile applications or ATMs, and expanding their scope as benefits materialize.

Additionally, partnerships with technology providers and access to government grants or tax incentives can help reduce the financial burden of implementation. These strategies allow financial institutions to take advantage of the benefits of biometrics without compromising financial stability.

5. Case Studies and Best Practices

5.1. BBVA Bank

The financial institution, BBVA, was a pioneer in Spain by introducing the possibility of signing banking transactions using biometric systems through its mobile application. This functionality was achieved through collaboration with Veridas, a leading international digital identity verification company. The integration of biometric technologies, namely, facial and fingerprint recognition, has raised the level of security for customers and protected them against potential fraud, and streamlined operational processes, offering a more efficient and reliable user experience. To ensure that this innovation responds to the real needs of users, BBVA conducted a comprehensive study based on in-depth interviews with customers and non-customers in Spain. This analysis allowed us to identify preferences and perceptions regarding access and security methods. Among the key findings, the ideal signature method should balance convenience and security, ensuring that the customer receives clear confirmation of the transaction. This customer-centric approach underscores the importance of tailoring technology to meet end-user expectations.

In its efforts to lead innovation in digital authentication, BBVA became the first bank in Spain to integrate the Fast Identity Online (FIDO) standard into its mobile application. This standard is internationally recognized as one of the most advanced solutions for online authentication, eliminating the need to enter digits or characters when accessing an application. With FIDO, online security is strengthened by a system that combines biometric authentication with advanced encryption technologies. This integration improves the protection of customer data and speeds up and simplifies access, enhancing the user experience. This innovative functionality allows users to interact seamlessly with digital banking without compromising security. This technology enhances customer security by combining facial and fingerprint biometrics, along with the FIDO standard, significantly minimizing the risks of fraud and identity theft, and providing a streamlined experience where customers can bank quickly and conveniently, without the need to remember complex passwords. Furthermore, by integrating these advanced technologies, BBVA reinforces its reputation as an innovative financial institution that focuses on protecting its users. Figure 2 shows details of BBVA's integrated system.

Figure 2: BBVA's integrated system. *Source:* BBVA website (2024).

5.2. EVO Bank

EVO Bank has integrated into its digital ecosystem an innovative biometric technology based on voice recognition, called "EVO VoiceID", which allows customers to make immediate transfers, validate banking transactions and access their mobile and online banking without the need for passwords. This solution has been made possible through partnerships with Nuance Communications, a leader in voice biometrics and conversational AI technologies.

The "EVO VoiceID" system is based on the creation of a unique biometric fingerprint for each customer, using the ability to analyze thousands of physiological characteristics of the voice. These include pitch, cadence, pauses, and other individual nuances that make the voice a highly secure and personalized identifier. This approach ensures accurate authentication and safeguards users against threats like theft, impersonation, or identity fraud.

This technology provides a superior level of protection, as it is virtually impossible to replicate, even with voice recordings or impersonations. Customers can also access their private digital banking environment using only their voices, eliminating the need to remember complex passwords or use additional devices. Owing to integrated conversational AI, the system validates the user's identity and facilitates smooth and natural interaction in transactions and banking. Figure 3 shows details of EVO Bank VoiceID system.

Figure 3: EVO Bank VoiceID system. *Source:* Evo Bank website (2024).

5.3. Japan Seven Bank

The system implemented by Seven Bank uses advanced facial recognition technology to verify the customer's identity before allowing access to the ATM. This process ensures that only the legitimate owner of the card can perform transactions, minimizing the risks associated with the fraudulent use of bank cards.

Facial recognition complements traditional authentication methods, namely, PIN codes, creating a multifactor security system that significantly hinders attempts at fraud or unauthorized access. Additionally, by automating the biometric verification process, reliance on physical items, including cards is reduced, thereby simplifying transactions for users.

This service has numerous advantages for customers. By verifying the customer's identity through unique biometric features, Seven Bank reinforces protection against identity theft and the unauthorized use of cards. Facial recognition also enables faster and more efficient transactions, eliminating the need for additional steps, namely, entering a PIN or using auxiliary devices. Finally, this system is especially useful for customers who may have difficulty remembering passwords or handling physical cards, thus making the service more inclusive.

5.4. Bradesco Bank

Banco Bradesco, one of Brazil's largest financial institutions, has implemented innovative biometric technology based on palm vein pattern recognition at its ATMs. This advanced authentication system offers a unique, secure, and highly reliable method of verifying the identity of users, marking a milestone in banking security in the region.

The palm-vein authentication system uses infrared technology to scan the subcutaneous vein pattern of the palm of the hand, which is unique to each individual and remains unchanged throughout their lives, making it highly accurate and virtually impossible to forge. Unlike other biometric technologies, namely, fingerprints or facial recognition, palm vein scanning works independently of external factors such as skin lesions, dirt, or lighting conditions.

The implementation of this technology in Banco Bradesco represents a significant advancement in the fight against bank fraud. It reinforces user authentication, ensuring that only authorized individuals can access their accounts and perform transactions. It protects customer assets and improves confidence in the services offered by the institution.

As in previous cases, this new biometric identification service has numerous advantages for customers. The palm vein pattern is externally invisible, making it difficult to copy or counterfeit. This provides an advanced layer of security over traditional authentication methods. Additionally, authentication is performed without direct contact, which improves hygiene and convenience, especially in high-traffic environments, like ATMs. Finally, users can complete their transactions quickly and seamlessly, eliminating reliance on physical cards or passwords.

5.5. Qatar National Bank (QNB)

QNB has implemented iris recognition technology in its ATMs, positioning itself at the forefront of innovation in banking security. This advanced functionality offers a higher level of protection by ensuring that only authorized users can access their accounts, adding an additional layer of security against potential threats.

Iris recognition is based on the identification of patterns in the colored part of the human eye that are unique to each individual and remain unchanged throughout life. Using specialized infrared cameras, this technology captures and analyzes these patterns with high accuracy, thereby ensuring virtually foolproof authentication.

Compared to other biometric methods like fingerprints and facial recognition, iris scanning stands out for its exceptional level of accuracy. Additionally, its non-invasive nature enables authentication without the need for physical contact, enhancing user experience and ensuring high standards of hygiene, which are especially relevant in public environments like ATMs.

Iris recognition offers superior security against fraud and unauthorized access because the iris pattern is extremely difficult to replicate or forge. Moreover, customers can authenticate their identity efficiently without the need to remember passwords or carry physical cards, thereby streamlining the transaction process. This biometric system also reinforces the perception of security in banking services, thereby increasing customer confidence in the institution.

6. Conclusions

6.1. Main Findings

This analysis demonstrates how biometric technologies have revolutionized the banking industry, establishing themselves as key tools to address critical security and customer experience challenges.

By leveraging unique physical and behavioral characteristics, biometrics have enabled a change in basic assumptions in user authentication, drastically reducing the vulnerabilities inherent in traditional methods. Technologies explored include facial, fingerprint, palm vein, iris, voice, retina recognition, and emerging approaches, namely, behavioral biometrics.

The impact of these technologies is reflected in improved security and optimization of customer experience. Case studies from banks, namely, BBVA, EVO Bank, Seven Bank, Banco Bradesco, and QNB have revealed how the strategic implementation of biometric solutions has improved transaction times, ensured transaction integrity, and generated a perception of innovation among customers. For example, the incorporation of facial recognition in BBVA's ATMs or palm vein technology in Banco Bradesco has redefined authentication standards in their respective regions.

However, this analysis also identified significant challenges, including high upfront costs, technological limitations in diverse contexts, and growing concerns regarding the privacy of biometric data. These challenges underscore the need to adopt these technologies with well-planned strategies that balance the benefits with the technical and ethical complexities involved.

The findings confirm that biometrics is an effective solution to current security issues in banking and represents an opportunity to transform customer interaction with financial services, establishing itself as a fundamental pillar for future banking.

6.2 Implications for Security and Customer Experience

Biometric Technologies in Banking Security and Compliance with International Regulations

Biometric technologies have significantly increased the level of security in the banking industry. By relying on unique and non-transferable characteristics of users, namely, iris patterns, fingerprints, or voice traits, these technologies virtually eliminate the risks associated with credential theft, unauthorized access, and fraud like phishing or spoofing. This accurate and strong authentication capability is further enhanced by the integration of advanced techniques, including liveness detection, which prevents impersonation through recordings or spoofing.

Biometric technologies enable financial institutions to comply with increasingly stringent international regulations, namely, GDPR, KYC, and AML regulations. These tools ensure the integrity of transactions and protect the reputation of institutions by demonstrating an initiative-taking commitment to security and compliance.

Challenges of Managing Biometric Data

These solutions are not without challenges. The management of biometric data raises serious privacy and security concerns. Institutions must implement advanced encryption, anonymization, and access control measures to protect sensitive information and ensure that systems are resistant to cyberattacks. Compliance with these standards is essential from a legal perspective, and to build and maintain customer trust.

Enhancing Customer Experience with Biometrics

Beyond security, biometric technologies have transformed customer interaction with financial services, offering a more streamlined, intuitive, and personalized experience. By eliminating the need to remember complex passwords or use additional devices, these technologies significantly simplify banking processes. Customers can perform transactions, access their accounts, and validate operations quickly and smoothly, improving their perception of efficiency and modernity in the institution.

Convenience and Relevance in a Post-pandemic Era

Recent studies have shown that customers perceive biometrics as a secure solution and a tool that increases their convenience. For example, facial recognition and voice biometrics enable contactless interaction, which is convenient and relevant in a post-pandemic context where hygiene measures are a priority. This customer-centric approach fosters greater loyalty and satisfaction, positioning institutions that adopt these technologies as leaders in innovation.

Addressing Privacy Concerns and Building Trust

However, it is crucial to address users' concerns about privacy and handling of their biometric data. Transparency in the collection and use of information, combined with effective communication about the benefits and safeguards in place, is critical for mitigating these concerns. By striking a balance between security and convenience, financial institutions can solidify their positions as trusted and visionary players in the marketplace.

Future Implications of Biometrics in Banking

Overall, the implications of biometric technologies in banking go beyond security and customer experience, transforming the way financial institutions interact with their users and respond to digital environmental challenges. These technologies solve immediate problems, and lay the foundation for a more secure, efficient, and customer-centric future.

6.3. Recommendations for Future Research and Adoption in the Banking Industry

The evolution of biometric technologies in banking has opened a promising horizon for both security enhancement and customer experience. To maximize the impact, it is essential to address certain critical aspects through future research and implementation strategies. First, it is imperative to explore the potential of multimodal biometric technologies. Combining methods, namely, facial recognition, iris recognition, and behavioral biometrics, can provide more robust authentication by minimizing errors and maximizing accuracy. This multimodal approach improves security and allows systems to be tailored to the specific needs of different customer segments. Research should focus on optimizing the integration of these technologies, ensuring that their implementation does not create friction in the user experience. However, the development of interoperability standards is a crucial issue that requires immediate attention. The absence of unified frameworks for integrating biometric systems on different platforms limits their widespread adoption. The creation of common standards would facilitate compatibility between technologies, allowing financial institutions to operate in global environments and reduce the costs associated with technological fragmentation. This would also foster collaboration between banks, regulators, and technology providers, paving the way for more scalable and accessible solutions.

The incorporation of AI and ML into biometric systems represents another area with significant potential. These technologies can detect anomalies and fraud patterns in real time, thereby enhancing security measures. However, comprehensive studies are necessary to evaluate how these tools can be integrated efficiently to mitigate algorithmic biases that may affect the accuracy and fairness of the system. This approach may be instrumental in addressing dynamic and adaptive fraud challenges in the financial sector. Furthermore, the risks and potential applications of AI by malicious actors (for instance, the utilization of generative AI tools in voice impersonation) warrant serious consideration, particularly in the light of rapid evolution of these systems.

A fundamental aspect that cannot be overlooked is the privacy of biometric data. Ensuring the protection of sensitive information is a priority in any implementation strategy. Adopting the Privacy by Design principle is essential to comply with global regulations, like the (GDPR), and to gain customer trust. Future research should focus on developing solutions that minimize data collection, use advanced anonymization techniques, and ensure end-to-end encryption of stored information.

Additionally, biometrics have enormous potential to improve financial inclusion, especially in regions with limited infrastructure or underserved populations. Thus, it is crucial to design low-cost technologies that can be implemented in emerging markets to ensure that these solutions are accessible to all layers of society. The combination of mobile biometrics and digital services can be a powerful tool for extending access to basic financial services in remote communities.

Finally, it is necessary to comprehensively measure the financial and social impacts of these technologies. Studies that analyze return on investment and social benefits, like reduced fraud and increased customer confidence, are essential to justify investments and guide smaller institutions in adopting these solutions. This type of analysis must be complemented by effective communication strategies that educate customers about the benefits of biometrics, address their concerns, and encourage greater acceptance of these technologies.

References

Al-Debei, M.M., Hujran, O., & Al-Adwan, A.S. (2024). Net valence analysis of iris recognition technology-based FinTech. *Financial Innovation*, 10(1), 59.

Ali, A., Baghel, V.S., & Prakash, S. (2023). A novel technique for fingerprint template security in biometric authentication systems. *The Visual Computer*, 39(12), 6249–6263.

Alrawili, R., AlQahtani, A.A.S., & Khan, M.K. (2024). Comprehensive survey: Biometric user authentication application, evaluation, and discussion. *Computers and Electrical Engineering*, 119, 109485.

Aratek. (2023). Cómo la biometría ayuda a mejorar la inclusión financiera en la banca. Available at https://www.aratek.co/es/news/biometrics-in-banking-how-biometrics-make-better-financial-inclusion.

Asl, M.G., Rashidi, M.M., & Abad, S.A.H.E. (2021). Emerging digital economy companies and leading cryptocurrencies: Insights from blockchain-based technology companies. *Journal of Enterprise Information Management*, 34(5), 1506–1550. https://doi.org/10.1108/JEIM-08-2020-0348.

Awadallah, A., Eledlebi, K., Zemerly, J., Puthal, D., Damiani, E., Taha, K., ... & Yeun, C.Y. (2024). Artificial intelligence-based cybersecurity for the metaverse: Research challenges and opportunities. *IEEE Communications Surveys & Tutorials*, 1–45.

Ayeswarya, S., & Singh, K.J. (2024). A comprehensive review on secure biometric-based continuous authentication and user profiling. *IEEE Access*, 12, 82996–83021.

Barkadehi, M.H., Nilashi, M., Ibrahim, O., Fardi, A.Z., & Samad, S. (2018). Authentication systems: A literature review and classification. *Telematics and Informatics*, 35(5), 1491–1511.

Basare, A., Bhojak, D., & Solanki, R. (2023). Biometric authentication system. *International Journal for Research in Applied Science and Engineering Technology*, 11(6), 3232–3238. https://doi.org/10.22214/ijraset.2023.54246.

BBVA (2024). La 'app' móvil de BBVA Paraguay mejora la experiencia digital de los clientes. Available at https://www.bbva.com/es/innovacion/app-movil-bbva-paraguay-mejora-experiencia-digital-clientes/

Bhartiya, N., Jangid, N., & Jannu, S. (2018). Biometric authentication systems: Security concerns and solutions. In: *2018 3rd International Conference for Convergence in Technology (I2CT)*, 1–6. https://doi.org/10.1109/i2ct.2018.8529435.

Boudjellal, S.E. (2024). Biometric Security System: Unimodal Identification Using Finger Veins (Doctoral dissertation). http://dspace.univ-setif.dz:8888/ispuui/handle/123456789/4371.

Choquehuanca-Chuctaya, C., & Arroyo-Paz, A. (2024, August). The security of biometric data in devices with cancellable biometrics technology: A systematic review of the literature. In: *2024 4th International Conference on Emerging Smart Technologies and Applications (eSmarTA)* (pp. 1–9). IEEE.

Cross, M.P., Acevedo, A.M., & Hunter, J.F. (2023). A critique of automated approaches to code facial expressions: What do researchers need to know? *Affective Science*, 4(3), 500–505.

Deloitte. (2021). Digital Banking Maturity 2020. Available at https://www2.deloitte.com/content/dam/Deloitte/ce/Documents/financial-services/ce-digital-banking-maturity-2020.pdf.

Dharavath, K., Talukdar, F.A., & Laskar, R.H. (2013). Study on biometric authentication systems, challenges, and future trends: A review. In: *2013 IEEE International Conference on Computational Intelligence and Computing Research*, 1–7. https://doi.org/10.1109/iccic.2013.6724278.

Dobbie, S. (2020). Challenge of biometric security for banks. *Biometric Technology Today*, 2020(3), 5–7. https://doi.org/10.1016/s0969-4765(20)30037-0.

Elhag, S., & Alharbi, R.S.S. (2024). Behavioral biometrics – A case study of digital banking security in Saudi. In: *AI Applications in Cyber Security and*

Communication Networks: Proceedings of Ninth International Conference on Cyber Security, Privacy in Communication Networks (*ICCS 2023*) (Vol. 1032, p. 39). Springer Nature.

European Union. (2016). General Data Protection Regulation (GDPR). Available at https://gdpr-info.eu/

EVO. (2024). Más Voice Bank que nunca: Ahora con EVO Voice ID, tu contraseña será tu voz. Available at https://blog.evobanco.com/evo-voice-id-contrasena-voz/

Fatima, A. (2011). E-banking security issues: Is there a solution in biometrics? *Journal of Internet Banking and Commerce*, 16(2), 1.

Goode Intelligence. (2019). Biometrics for Financial Services: Market and Technology Analysis, Adoption Strategies, and Forecasts 2018–2023. Available at https://www.goodeintelligence.com/report/biometrics-for-financial-services-market-technology-analysis-adoption-strategies-forecasts-2018-2023/

Goode Intelligence. (2020). Biometrics for Financial Services: Market and Technology Analysis, Adoption Strategies, and Forecasts 2020–2025. Available at https://www.goodeintelligence.com/report/biometrics-for-financial-services-market-technology-analysis-adoption-strategies-forecasts-2020-2025/

Hassan, W., & Sabahat, N. (2024). Towards secure identification: A comparative analysis of biometric authentication techniques. *VFAST Transactions on Software Engineering*, 12(1), 105–120.

IBM Security. (2020). Cost of a Data Breach Report 2020. Available at https://www.ibm.com/security/digital-assets/cost-data-breach-report/

Instituto Nacional de Seguridad. (2024). Biometría: amenazas, riesgos y vulnerabilidades. Available at https://www.incibe.es/empresas/blog/biometria-amenazas-riesgos-y-vulnerabilidades.

Iyawe, I.B. (2010). Towards implementing an efficient biometric authentication framework for nigeria customer banking systems. *Information Manager* (*The*), 10(1–2), 22–32.

Jain, A.K. (2007). *Handbook of Biometrics*. Springer Google Scholar, 2, 114–116.

Jain, A.K., Nandakumar, K., & Ross, A. (2016). 50 years of biometric research: Accomplishments, challenges, and opportunities. *Pattern Recognition Letters*, 79, 80–105.

Javelin Strategy & Research. (2020). 2020 Identity Fraud Study: Genesis of the Identity Fraud Crisis. Available at https://www.javelinstrategy.com/coverage-area/2020-identity-fraud-study-genesis-identity-fraud-crisis.

Juniper Research. (2021). Mobile Payment Authentication: Biometrics, Regulation, & Market Forecasts 2021–2025. Available at https://www.juniperresearch.com/researchstore/fintech-payments/mobile-payment-authentication-research-report.

Kantaros, E. (2023). Biometric security authentication: A cost-benefit review. *Cornell Undergraduate Economic Review*, p. 56. New York: UNICEF.

Kataria, A.N., Adhyaru, D.M., Sharma, A.K., & Zaveri, T.H. (2013). A survey of automated biometric authentication techniques. In: *2013 Nirma University International Conference on Engineering* (*NUiCONE*), (pp. 1–6). https://doi.org/10.1109/nuicone.2013.6780190.

Khan, H.U., Malik, M.Z., Nazir, S., & Khan, F. (2023). Utilizing bio metric system for enhancing cyber security in banking sector: A systematic analysis. *IEEE Access*, 80181–80198.

Khare, P., Arora, S., & Gupta, S. (2024, August). Artificial intelligence-based biometric authentication systems for facial recognition and identification. In: *2024 International Conference on Electrical Electronics and Computing Technologies* (*ICEECT*) (Vol. 1, pp. 1–7). IEEE.

Kuznetsov, O., Zakharov, D., & Frontoni, E. (2024). Deep learning-based biometric cryptographic key generation with post-quantum security. *Multimedia Tools and Applications*, 83(19), 56909–56938.

Liébana-Cabanillas, F., Kalinic, Z., Muñoz-Leiva, F., & Higueras-Castillo, E. (2024). Biometric m-payment systems: A multi-analytical approach to determining use intention. *Information & Management*, 61(2), 103907.

Lucia, C., Zhiwei, G., & Michele, N. (2023). Biometrics for Industry 4.0: A survey of recent applications. *Journal of Ambient Intelligence and Humanized Computing*, 14(8), 11239–11261.

Marani, M., Soltani, M., Bahadori, M., Soleimani, M., & Moshayedi, A. (2023). The role of biometric in banking: A review. *EAI Endorsed Transactions on AI and Robotics*, 2(1), 1–15.

Mohammed, S.M., Mashhadani, S., & Ali, O. (2024). Human biometric identification: Application and evaluation. *IJECS*, 6(2), 131–152.

Mubeen, M. (2024). Biometric authentication: Past, present, and future perspectives. *International Journal of Innovative Research in Technology and Science*, 12(2), 351–362.

Nakisa, B., Ansarizadeh, F., Oommen, P., & Kumar, R. (2023). Using an extended technology acceptance model to investigate facial authentication. *Telematics and Informatics Reports*, 12, 100099.

Normalini, M.K., & Ramayah, T. (2012). Biometrics technologies implementation in Internet banking reduces security issues? *Procedia-Social and Behavioral Sciences*, 65, 364–369.

Piotrowska, A. (2024). Determinants of consumer adoption of biometric technologies in mobile financial applications. *Economics and Business Review*, 10(1), 81–100.

Piotrowska, A.I., Polasik, M., & Piotrowski, D. (2017). Prospects for the application of biometrics in the Polish banking sector. Equilibrium. *Quarterly Journal of Economics and Economic Policy*, 12(3), 501–518.

Ponemon Institute. (2019). The Impact of Data Breaches on Reputation & Share Value. Available at https://www.centrify.com/media/4737054/ponemon_data_breach_impact_study.pdf

Poullet, Y. (2023). The data protection impact assessment or rather the privacy impact assessment, a revolution with a future in the age of artificial intelligence? In: *Artificial Intelligence Law: Between Sectoral Rules and Comprehensive Regime: Comparative Law* (pp. 627–649). Bruylant.

Rashidi, F.U., Mohsini, M.H., & Mega, B. (2024). A framework for security improvement on usage of mobile money application based on iris biometric authentication method. *Information Security Journal: A Global Perspective*, 1–13.

Revolution Banking. (2021). Biometría en banca: cada cliente es único. Available at https://www.revolutionbanking.es/biometria-banca/

Romero, C. (2018). Seis Retos en la Implementación de Sistemas Biométricos. Fintechgracion. Available at https://www.fintechgracion.com/2018/02/seis-retos-la-implementacion-sistemas-biometricos/

Ross, A., Nandakumar, K., & Jain, A.K. (2008). Introduction to multibiometrics. In: *Handbook of Biometrics*, 271–292.

Ryu, R., Yeom, S., Herbert, D., & Dermoudy, J. (2023). The design and evaluation of adaptive biometric authentication systems: Current status, challenges, and future direction. *ICT Express*, 9(6), 1183–1197.

Singh, N., Agrawal, A., & Khan, R.A. (2018). Voice biometric: A technology for voice based authentication. *Advanced Science, Engineering, and Medicine*, 10(7–8), 754–759.

Smith-Creasey, M. (2023). Continuous authentication. In: *Continuous Biometric Authentication Systems: An Overview* (pp. 35–72). Cham: Springer International Publishing.

Soffid. (2023). El Futuro de la Autenticación Biométrica: Desafíos y Perspectivas. Available at https://www.soffid.com/es/el-futuro-de-la-autenticacion-biometrica-desafios-y-perspectivas/

Sudhakar, K., Raju, N.P.K., Praveen, V., & Hithin, M. (2024). Advanced authentication system: A secure authentication methodology incorporating face and text passwords, image password grids, sensitive information retrieval, and login activity tracking with efficient database connection. *International Journal of Research in Engineering, Science, and Management*, 7(2), 89–94.

Sumalatha, U., Prakasha, K.K., Prabhu, S., & Nayak, V.C. (2024). A comprehensive review of unimodal and multimodal fingerprint biometric authentication systems: Fusion, attacks, and template protection. *IEEE Access*, 64300-64334.

Thakur, R., Kumar, S., Singh, S.K., Singla, K., Sharma, S.K., & Arya, V. (2024). Cyber synergy: Unlocking the potential use of biometric systems and multimedia forensics in cybercrime investigations. In: *Digital Forensics and Cyber Crime Investigation*. CRC Press.

Tripathi, D.R., & Nishad, D.K. (2020). Biometric authentication systems: A survey. *Turkish Journal of Computer and Mathematics Education (TURCOMAT)*, 11(3), 2878–2884. https://doi.org/10.61841/turcomat.v11i3.14653.

Venkatraman, S., & Delpachitra, I. (2008). Biometrics in banking security: A case study. *Information Management & Computer Security*, 16(4), 415–430. https://doi.org/10.1108/09685220810908813.

Venugopal, H., & Viswanath, N. (2016, November). A robust and secure authentication mechanism in online banking. In: *2016 Online International Conference on Green Engineering and Technologies (IC-GET)* (pp. 1–3). IEEE.

Visa. (2020). Consumer Payment Attitudes Study. Available at https://www.visa.com.sg/about-visa/newsroom/press-releases/consumer-payment-attitudes-study-2020.html.

Wang, C., Tang, H., Zhu, H., Zheng, J., & Jiang, C. (2024). Behavioral authentication for security and safety. *Security and Safety*, 3, 2024003.

Yang, W., Wang, S., Sahri, N.M., Karie, N.M., Ahmed, M., & Valli, C. (2021). Biometrics for internet-of-things security: A review. *Sensors*, 21(18), 6163.

5

Exploring the Integration of Biometric Technologies in the Macedonian Banking Sector: A Qualitative Study

Jana Prodanova[1*(0000-0001-6381-3621)], **Anita Ciunova-Shuleska**[2(0000-0001-6402-7630)], **Nikolina Palamidovska-Sterjadovska**[2(0000-0002-2926-5649)] **and Ljupco Kocarev**[1(0000-0001-7243-7074)]

[1] Macedonian Academy of Sciences and Arts, Blvd. Krste Misirkov, 2, 1000 Skopje, R. Macedonia

[2] Faculty of Economics-Skopje, Ss. Cyril and Methodius University in Skopje, Blvd. Goce Delchev 9V, 1000 Skopje, R. Macedonia

1. Introduction

Technological advancements and evolving customer expectations have brought considerable changes in the banking industry (Broby, 2021; Ciunova-Shuleska et al., 2022; Vivekanandan and Jayasena, 2012). Technologies like NFC (Near Field Communication) and e-wallets are becoming popular, leading to decreased cash usage and fewer traditional debit and credit card transactions. The predictions are that the number of cashless payments worldwide will increase by more than 80% by 2025 and triple by 2030, compared to 2020 (PwC, 2021).

This trend highlights the need for banks to adapt to evolving payment preferences, thereby improving customer experience (Liébana-Cabanillas et al., 2021; Ciunova-Shuleska et al., 2022; Palamidovska-Sterjadovska et al., 2024). However, security and privacy remain important concerns

*Corresponding author: jprodanova@manu.edu.mk

for financial institutions (Banga & Pillai, 2021). A highly effective method for securing customer-bank interactions is the implementation of biometric technologies, which with the growth of financial technology utilizing mobile devices and apps, have become essential for improving the identification and security of these applications (Wang, 2021). Many banks have successfully integrated biometric solutions into their banking services (e.g., ATM cash withdrawals and mobile banking apps) to prevent fraud and identity theft (Svilar & Zupančič, 2016).

O'Neil and Burt (2017) argue that the rapid adoption of mobile banking and mobile commerce is largely due to the integration of biometric technology into smartphones. The rise in smartphone users with biometric features contributes to the biometrics market growth as these devices allow secure and convenient access to services like mobile payments (Reportlinker, 2024). Therefore, over one-third of respondents across industries worldwide plan to adopt passwordless authentication in the next one to three years, whereas 39% of respondents from the financial services industry claimed that they would most likely adopt passwordless authentication soon, opening the opportunity for biometrics adoption (Statista, 2023).

Despite the promising trends in digital banking, several challenges hinder the widespread adoption of biometric technologies related to technological development, costs, management challenges, and legal and ethical considerations (Venkatraman & Delpachitra, 2008). While understanding the usability of biometrics is essential, addressing the challenges surrounding biometrics' implementation is critical for overcoming barriers to their broader use in various sectors, including banking. However, only a few studies have investigated biometrics' application within the banking industry (Toledano et al., 2006; ChitraKiran et al., 2017; Wang, 2021; Soto-Beltrán et al., 2022), primarily focusing on customer perspectives. Thus, there is a significant gap in studies evaluating these technologies from user-centered, business-centered, and IT-centered perspectives.

The present study aims to fill these gaps by exploring the opportunities and challenges of implementing biometrics in banking from an organizational perspective, specifically from the standpoint of banks and IT experts who develop banking applications. Furthermore, while a few studies discussed biometrics in the context of migration management in Balkan countries (Ivanović & Ivanović, 2018; Vukomanović, 2021), there is an absence of research explicitly examining the application of biometrics in the banking sector (Odorović et al., 2020). Mobile banking is rapidly expanding in Macedonia, with 54.2% of individual payments conducted via mobile apps (NBR NM, 2022). In the first quarter of 2021, mobile payments increased by 99% compared to the previous year, which is twice the growth rate recorded during the same period in 2020 (NBR NM, 2021).

This study will be the first qualitative research to use a combination of thematic, sentiment, and content analysis to examine the responses from bank managers and IT experts responsible for designing and deploying banking technologies, providing a comprehensive understanding of their perspectives on the implementation and future development of biometrics in the sector. Bearing in mind the aforementioned, the main research questions that this study, in Macedonia, wants to address are:

RQ1: What is the current landscape of biometrics implementation in banking services?

RQ2: What are the challenges of implementing biometrics in banking services?

RQ3: What is the emotional tone (sentiment) of the discussed topics regarding biometrics?

RQ4: What are the expectations regarding the future of biometrics in banking services?

2. Literature Review

Biometric solutions encompass identity authentication based on physical traits (face and fingerprint recognition, fingerprint, hand geometry, palm print, or iris) and behavioral traits (signature, voice, gait, or keystroke dynamics) (Jain et al., 2004; Biometrics Research Group, 2009; Sheena & Sheena, 2014). By using all these biometric solutions, banks can verify the identity of account holders during transactions, offering increased security compared to conventional authentication methods (Clark, 2021). However, not all these solutions are suitable and preferable for large-scale biometrics applications for civilian usage, particularly in the banking industry (Wang, 2021). Among four alternative authentication methods for mobile banking applications (fingerprint, PIN, token, and card reader), fingerprint authentication was perceived as the most usable and secure method (Kruzikova et al., 2022; Liébana-Cabanillas et al., 2024).

Banks today are impacted by a complex array of techno-economic, legal, ethical, and operational factors that play a critical role in their efforts to provide adequate security solutions while considering various constraints, expectations, standards, and legal requirements from both customer and business perspectives. Concerning customer perspective, few studies have been conducted on the acceptance of biometrics in banking, but they have recently surfaced. Wang (2021) applied an extended total addressable market (TAM) to examine user acceptance of biometric identification in FinTech applications, proposing face and voice recognition as the most preferred identification modules. Based on TAM, Soto-Beltrán et al. (2022) suggested that users are willing to accept biometrics in banking if perceived as useful and risk-free, even though end-users may resist biometric technologies due security, risk, and invasiveness concerns (Pons & Polak, 2008), often driven by fears of privacy violations.

Security and privacy are the crucial priorities in bank digitalization for maintaining customer trust, reputation, and successful operations (Martínez-Navalón et al., 2023). Biometric technologies must ensure a balance between privacy and security (Jain et al., 2006), even though the security benefits have far outweighed the concerns relating to privacy (Jain & Kumar, 2010). Technological advancements offer significant opportunities for banks and other companies to enhance data protection and security, namely, passwordless authentication and biometrics technology (Parmar et al., 2022). Tassabehji and Kamala (2009) empirically proved that users' perceptions of biometrics security positively influence their attitude and intention to use biometric banking. They found that perceived e-banking security has a substantial influence on behavioral intention to adopt biometric banking than its perceived usefulness.

With the increasing prevalence of transaction fraud and identity theft, the banking and finance sector faces a critical need for highly secure identification and personal verification systems, making biometric technology an almost perfect solution to address these security challenges (Normalini & Ramayah, 2012). Banks should prioritize investing in awareness programs to educate customers about the risks associated with online banking and the measures they can take to protect themselves (Telo, 2023). Beyond security and privacy, biometrics in mobile banking significantly enhances the user experience, ensuring that financial transactions are convenient and simplified (Suganya et al., 2012; Clark, 2021).

Beyond the customer perspective, Venkatraman and Delpachitra (2008) analyzed biometrics in banking from a business perspective and identified four groups of biometrics success factors: technology, monetary, management, and legal and ethical factors. They found that establishing a sustainable biometrics security infrastructure, and addressing legal and ethical issues along with good management and leadership support focusing on managing the biometrics' cost-effectiveness and its benefits are crucial for the successful implementation and utilization of biometrics in banking organizations. Recently, information security policies and regulatory frameworks have evolved substantially (Karger et al., 2000; Tanwar et al., 2019; Smith & Urbas, 2021). According to Lovisotto et al. (2017, p. 6), regulatory considerations are gaining importance as industry professionals emphasize that they frequently "delay the design process and significantly slow down decision-making due to the heightened risks involved".

Most of the discussed studies focus on biometrics from end-user perspectives, although ultimately, the users who operate the technology in the organization (managers and employees) have a crucial role in the outcome. Additionally, having the right technology and established rules and processes are essential for success. Therefore, this research aimed to gain a deeper understanding of the status of biometrics implementation in the banking sector, potential benefits and challenges associated with biometric technology application, and user readiness and trust in biometrics, by gathering insights from key stakeholders, i.e., bank managers and IT professionals responsible for developing financial software.

3. Methodology

This study employed a qualitative research design to explore the integration of biometric technologies within the Macedonian banking sector. This approach was chosen due to its suitability for capturing the nuanced perspectives and experiences of individuals directly involved in implementing and managing these technologies (Neuman, 2014). Figure 1 depicts the stages of the qualitative analysis process.

The participants in this study were selected using purposive sampling to ensure that those interviewed had relevant experience with biometric technologies in the banking sector. A total of 11 individuals participated, including bank managers and IT specialists from various Macedonian banks. They were chosen based on their decision-making roles and involvement in deploying biometric systems within their respective institutions. This approach ensured that the data collected

Figure 1: Stages of qualitative analysis.

would provide valuable insights into the banking industry's strategic and operational aspects of biometric technology implementation.

Data were collected through semi-structured interviews, which allowed for a focused yet flexible exploration of the participants' views on biometric technologies. The interviews, lasting 30 to 45 minutes, were conducted in person or online, depending on the participants' preferences and availability. The open-ended interview questions let participants share their thoughts on the convenience, impact, and potential risks associated with biometric technologies in banking. This allowed participants to freely express their thoughts and experiences, enriching the data with diverse perspectives.

The data were analyzed using a combination of thematic, sentiment, and content analysis facilitated by NVivo software. Thematic analysis was employed to identify recurring themes and patterns within the interview data, following the approach outlined by Braun and Clarke (2006). Thematic analysis was conducted in several stages: initially, transcripts were reviewed to familiarize the researchers with the content, followed by generating initial codes, which were then grouped into broader themes that captured recurring ideas and sentiments across the interviews. These themes were further refined and defined to ensure they accurately represented the data. By using an interpretive approach that does not require quantifiable inter-coder reliability, researchers can better understand the phenomenon they are looking at (Braun & Clarke, 2021).

Additionally, sentiment analysis was performed to determine the overall emotional tone of the responses, categorizing them as positive, negative, or neutral. Content analysis was used to pinpoint specific issues or concerns raised by the participants, particularly regarding privacy, user preferences, and future possibilities for biometric technologies in banking.

4. Results

The analysis yielded several key findings, categorized into thematic analysis, sentiment analysis, and content analysis, where direct quotes and comparative analysis reveal further conclusions about the current situation of biometrics usage in the banking sector. Moreover, the challenges banks face when implementing biometric technologies were likewise explored. Additionally, strategies for addressing privacy concerns were offered based on the participants' suggestions.

4.1. Thematic Analysis

The analysis revealed five primary themes: implementation status of biometric technologies, user readiness and trust, potential applications of biometrics, regulatory environment, and challenges in implementation.

The study found that the *adoption of biometric technologies* varied significantly among the banks. Some had implemented advanced systems like fingerprint and facial recognition, particularly in mobile banking applications and ATMs, while others were still in the exploratory stages. The primary motivations for adoption included enhancing security, improving customer experience, and increasing operational efficiency. The findings highlighted mixed perceptions regarding *user readiness and trust* in biometric systems. Some participants believed customers were generally ready to adopt these technologies, while others pointed out significant hesitancy due to a lack of awareness and understanding. The generational divide was also evident, with younger customers being more open to using biometrics than older customers, who may have concerns about privacy and data security. Participants identified a broad range of *potential applications for biometric technologies* within banking, including customer authentication for mobile banking and ATMs and more secure online payment verification processes. The potential for future expansion was noted, with participants suggesting that more advanced biometric solutions, like voice and retinal recognition, could become prevalent as technology evolves. The *regulatory environment* was a recurring theme, with participants highlighting the partial support from Macedonian regulations for using biometric technologies. However, the lack of a fully developed legal framework poses uncertainties and challenges for banks, particularly regarding compliance with data protection laws. Finally, the analysis revealed several *challenges* associated with implementing biometric technologies. Technical and cost barriers were frequently mentioned, particularly integrating biometric systems into existing banking infrastructure. Participants also emphasized the need for better customer education and trust-building efforts to address privacy and data security concerns.

The most frequent terms mentioned in the interviews to describe the identified themes are presented in Figure 2.

4.2. Sentiment Analysis

The sentiment analysis provided an overview of the emotional tone of the

Figure 2: Word frequency in the thematic analysis.

participants' responses. Overall, there was a predominance of positive sentiments, particularly regarding the perceived security benefits of biometric technologies. Participants highlighted these technologies' enhanced security, especially in preventing fraud and unauthorized access. A positive sentiment was associated with the convenience and efficiency of biometric systems, which were seen as making banking more user-friendly and competent. Figure 3 shows the terms that identified the most positive sentiment reflected in the interviews.

Figure 3: Word frequency of very positive sentiment.

There were also notable negative sentiments, mainly related to concerns about privacy and trust. Some participants expressed apprehensions about the security of biometric data, fearing potential misuse or breaches. Additionally, the complexity and cost of integrating biometric technologies were cited as significant barriers, contributing to negative sentiments about their implementation.

Figure 4 shows the summary of the terms describing the negative sentiment.

Figure 4: Word frequency of very negative sentiment.

4.3. Content Analysis

The content analysis identified specific issues raised by the participants, categorized into positive aspects, negative aspects, customer preferences, and future possibilities for biometric technology use in banks.

Participants frequently mentioned the increased security provided by biometric technologies for authentication in mobile banking and online transactions. The improved user experience, characterized by ease of use and faster transaction processes, made biometrics banking processes more convenient while providing robust security against fraud. A significant issue raised by participants in terms of the negative aspects of biometric technology was the lack of user trust and awareness, which can hinder its adoption in banking services. Privacy concerns were particularly prominent, with some participants pointing out that users may not fully understand how their biometric data is collected, stored, and used. High implementation costs were also a major concern, especially for smaller banks with limited resources.

Participants noted that customers generally prefer more straightforward biometric methods, namely, thumbprint recognition, over more complex options like facial or voice recognition. This preference for simplicity was attributed to familiarity and ease of use. There was a clear generational gap in customer preferences: younger customers being more inclined while older customers showed less interest. Lastly, the findings revealed optimism about the future potential of biometric technologies in the banking sector. Participants suggested that biometric technologies could be increasingly integrated into digital banking channels, like mobile apps and online platforms, to enhance security and user experience. There was also an expectation that as technology evolves, advanced biometric features will be implemented, offering greater security and convenience.

As part of the qualitative research, we aimed to identify specific challenges banks face when implementing biometric technologies. The results showed that many users are not fully aware of how biometric technologies work, leading to mistrust or reluctance to adopt these systems. Privacy concerns were prevalent, with users worrying about how their biometric data will be stored, used, and protected.

"Awareness and understanding are crucial. Many users still have reservations about privacy and data security, which makes it difficult for banks to promote biometric services effectively."

Compliance with regulations was also highlighted as a significant challenge. The legal framework in Macedonia is not fully developed or uniformly enforced, creating uncertainty for banks regarding data protection and privacy laws.

"The inconsistency in regulations or partial support from the legal system may lead to challenges in standardizing the implementation of biometric technologies across the banking sector."

Implementing biometric technologies requires significant upgrades to existing IT infrastructure, which can be costly and complex, especially for smaller banks.

"Integrating new technologies with existing systems is challenging and resource-intensive, particularly for smaller institutions."

The high initial investment required for biometric technologies is a significant barrier, especially for smaller banks. Additionally, banks must invest in training staff and educating customers on using biometric systems effectively.

"The costs of implementing and maintaining biometric systems are high, which is a major concern for smaller institutions with limited resources."

There is notable resistance from older demographics regarding the adoption of biometric technologies, often due to a lack of familiarity or comfort with new technologies, which can slow adoption. *"Older customers may be less willing to adopt biometric technologies due to a lack of familiarity or comfort with new technologies. This generational divide requires targeted efforts to educate and support these users."*

A comparative analysis was also carried out to distinguish different perspectives based on the participants' roles, experience levels, and the banking entity they represented. Participants from larger banks reported more advanced implementations of biometric technologies, including facial recognition and thumbprint authentication, whereas those from smaller banks or with less experience in biometric technologies had limited or no implementation, reflecting disparities in resources and expertise. Bank managers, who interacted more directly with customers, emphasized the need for education and trust-building efforts, particularly among older customers who were more resistant to adopting biometric technologies, whereas IT professionals focused more on the technical challenges and benefits of biometric systems. While there was a consensus on the high costs and technical challenges of implementing biometric technologies, participants from larger banks were more optimistic about overcoming these barriers. They saw greater potential for expanding biometric use in digital banking channels.

5. Discussion

This study highlights both the potential benefits and challenges of integrating biometric technologies in the Macedonian banking sector. Generally, biometric technologies are considered to offer significant advantages in terms of security and user experience. They provide seamless and efficient methods of authentication, reducing the likelihood of fraud and enhancing the overall customer experience (Clark, 2021; Wang, 2021). Nevertheless, their adoption is hindered by challenges related to user trust, regulatory uncertainties, and high implementation costs. Users, particularly those unfamiliar with biometric technology, may question its reliability, while banks face the difficulty of navigating an evolving regulatory landscape that lacks clear guidance on privacy and data protection issues. The results show that the substantial initial investment required for biometric technologies, combined with the ongoing costs of upgrading and implementing new technological advancements, create a significant barrier, particularly for smaller banks.

The analysis highlights the need for targeted education and trust-building efforts, especially among older customers, who may be less familiar with digital technologies and more skeptical about the security of biometrics. Additionally, more precise regulatory frameworks are essential to provide legal clarity and establish standards for the use of biometrics (Smith & Urbas, 2021), which would, in turn, support their broader adoption in banking. Despite these challenges, there is optimism about the future potential of biometrics, with opportunities for expanding their use in digital banking channels and adopting more advanced features as technology evolves.

By addressing privacy concerns and enhancing security measures, banks can build trust and encourage the adoption of biometric technologies, ultimately improving customer experience and operational efficiency in the financial services sector. Furthermore, collaboration between banks, technology providers, and regulators is crucial to developing robust solutions that meet both security standards and consumer expectations. As technological advancements continue to drive the development of more sophisticated biometric systems, the banking sector must remain agile, continuously adapting to the changing landscape to fully harness the benefits of biometric authentication. By doing this, banks can strengthen their competitive position in a market that prioritizes digital services while helping create a safer and more trustworthy financial system.

6. Academic and Managerial Implications

This study contributes to the existing literature on biometric technology in banking, specifically in the Macedonian context. The limited empirical research on biometrics in the Balkan region (Odorović et al., 2020; Vukomanović, 2021) further adds value to this study. The insights gained from interviews with experts provide a deeper understanding of the implementation and application of biometrics in banking, and the challenges and regulatory aspects. Aside from most of the previous studies focusing on the consumers' perspective of the factors for adapting and using new technologies (Wang, 2021; Soto-Beltrán et al., 2022), this study shed light on the banking representatives and IT sector experts. The obtained qualitative data offered a contextual understanding of the topic, incorporating the experts' perspectives and experiences.

The presented findings are in line with the four categories of biometrics success factors (technology, monetary, management, and legal and ethical factors) identified by Venkatraman and Delpachitra (2008). More precisely, the results suggest that the technology factors in terms of reliability and security of the biometric systems, along with their efficiency and long-term benefits, are crucial to implementing biometric technology and improving its effectiveness. This is in accordance with the claims by Normalini and Ramayah (2012) that implementing biometrics in Internet banking can largely contribute to reducing security issues, thus, ensuring a safe and secure environment for clients to conduct e-banking transactions via implementing biometrics should be a top priority for every bank. The study also addresses the importance of proper management practices in ensuring that the biometric systems are well-managed and used to their utmost capacity, which resonates with the aspects referring to the management factor identified by Venkatraman and Delpachitra (2008).

Additionally, the study considers the ethical and legal implications of applying biometric technology, covering topics like data security, privacy, and regulatory compliance. This is consistent with the previously elaborated that security and privacy are the most crucial and delicate issues in banking in the age of digitalization (Banga & Pillai, 2021), along with the increasing need to establish regulatory frameworks (Lovisotto et al., 2017).

Educating users and communicating transparently regarding the security and privacy of data is also essential to enhance user readiness and trust. In line with the findings by Telo (2023), education and awareness programs can increase security awareness and improve customers' ability to protect themselves against potential risks. However, this investment in education, even though necessary, can pose a challenge due to the resources required to bring these customers to an adequate level of understanding.

Based on the interview responses, participants suggested several strategies that banks can implement to address privacy concerns related to biometric data. Participants emphasized the importance of banks being transparent about collecting, storing, and using biometric data. Clear communication regarding data handling practices can help alleviate fears and misconceptions about privacy. Providing detailed information and engaging in educational campaigns can build customers' trust. Next, implementing robust security protocols, namely, multi-factor authentication and advanced encryption, can protect biometric data from unauthorized access and misuse. Regular security audits and assessments can help banks identify vulnerabilities and take proactive steps to address them.

Furthermore, ensuring compliance with data protection laws and regulations is crucial for addressing privacy concerns. Participants suggested that banks should regulate their practices with national and international standards to ensure that biometric data is handled responsibly and legally. This aligns with the findings by Lovisotto et al. (2017) that the regulatory aspect of biometrics in banking is becoming more concerning. Finally, adopting data minimization principles, where only the necessary biometric data is collected and retained for the shortest time required, can reduce privacy risks. Allowing customers control over their biometric data, such as opting out or deleting their information, can enhance trust and provide a sense of security.

7. Limitations and Future Research Lines

While our study aims to provide valuable insights, it has its limitations that suggest future research streams. The data was obtained through semi-structured interviews with experts involved in biometrics in the banking sector, and although their relevance is not questioned, the sample size may limit the reflection of the discussion to the broader population. The sample and the data are specific to the country's context, and some cultural issues may be expected. Future research should consider increasing and broadening the sample and thus the possibilities for generalizations of the conclusions. Also, cross-country studies would be beneficial to compare the specifics of different cultural landscapes.

Regarding the data analysis procedures, it can be noted that the flexibility of thematic analysis allowed us to capture valuable insights into the participants' discussion, but it could lead to a certain level of subjectivity and theoretical incoherence. However, the present exploratory study will be used as a valuable foundation for developing more conclusive future studies by applying alternative research methods. Quantitative methods should be considered in future studies to statistically test and validate the crucial concepts in biometrics in banking, as identified in this study.

References

Banga, L., & Pillai, S. (2021, July). Impact of behavioural biometrics on mobile banking system. *Journal of Physics: Conference Series*, 1964(6), 062109. https://doi.org/10.1088/1742-6596/1964/6/062109.

Biometrics Research Group. (2009). What is biometrics? Biometrics Research Group. https://biometrics.cse.msu.edu/info/index.html.

Braun, V., & Clarke, V. (2006). Using thematic analysis in psychology. *Qualitative Research in Psychology*, 3(2), 77–101. https://doi.org/10.1191/1478088706qp063oa.

Braun, V., & Clarke, V. (2021). Can I use TA? Should I use TA? Should I not use TA? Comparing reflexive thematic analysis and other pattern-based qualitative analytic approaches. *Counselling and Psychotherapy Research*, 21(1), 37–47. https://doi.org/10.1002/capr.12360.

Broby, D. (2021). Financial technology and the future of banking. *Financial Innovation*, 7(1), 47. https://doi.org/10.1186/s40854-021-00264-y.

ChitraKiran, N., Teja, B., Suresh, S., Krishna, B., Akarsh, S.M., & Yomas, J. (2017, May). A biometric based payment system by using payee and payer module. In: *2017 2nd IEEE International Conference on Recent Trends in Electronics, Information, & Communication Technology (RTEICT)*. 2252–2256. https://doi.org/10.1109/rteict.2017.8257001.

Ciunova-Shulcska, A., Palamidovska-Sterjadovska, N., & Prodanova, J. (2022). What drives m-banking clients to continue using m-banking services? *Journal of Business Research*, 139, 731–739. https://doi.org/10.1016/j.jbusres.2021.10.024.

Clark, R.A. (2021). *Consumers Perspectives on Using Biometric Technology with Mobile Banking* (Doctoral dissertation). Walden University ProQuest Dissertations & Theses.

Ivanović, A.R., & Ivanović, A.B. (2018). Biometric registration of migrants in the process of managing migrant crisis in Europe. *Zbornik Instituta za kriminološka i sociološka istraživanja*. 2, 271–284.

Jain, A.K., Ross, A., & Prabhakar, S. (2004). An introduction to biometric recognition. *IEEE Transactions on Circuits and Systems for Video Technology*, 14(1), 4–20.

Jain, A.K., Ross, A., & Pankanti, S. (2006). Biometrics: A tool for information security. *IEEE Transactions on Information Forensics and Security*, 1(2), 125–143. https://doi.org/10.1109/tifs.2006.873653.

Jain, A.K., & Kumar, A. (2010). Biometrics of next generation: An overview. *Second Generation Biometrics*, 12(1), 2–3.

Karger, P., Austel, V., & Toll, D. (2000). *A New Mandatory Security Policy Combining Secrecy and Integrity* (Technical Report RC 21717 (97406)). IBM Research Division.

Kruzikova, A., Knapova, L., Smahel, D., Dedkova, L., & Matyas, V. (2022). Usable and secure? User perception of four authentication methods for mobile banking. *Computers & Security*, 115, 102603. https://doi.org/10.1016/j.cose.2022.102603.

Liébana-Cabanillas, F., Kalinic, Z., Muñoz-Leiva, F., & Higueras-Castillo, E. (2024). Biometric m-payment systems: A multi-analytical approach to determining use intention. *Information & Management*, 61(2), 103907. https://doi.org/10.1016/j.im.2023.103907.

Liébana-Cabanillas, F., Singh, N., Kalinic, Z., & Carvajal-Trujillo, E. (2021). Examining the determinants of continuance intention to use and the moderating effect of the gender and age of users of NFC mobile payments: A multi-analytical approach. *Information Technology and Management*, 22(2), 133–161. https://doi.org/10.1007/s10799-021-00328-6

Lovisotto, G., Malik, R., Sluganovic, I., Roeschlin, M., Trueman, P., & Martinovic, I. (2017). *Mobile Biometrics in Financial Services: A Five Factor Framework*. University of Oxford, Oxford, UK.

Martínez-Navalón, J.G., Fernández-Fernández, M. & Alberto, F.P. (2023). Does privacy and ease of use influence user trust in digital banking applications in Spain and Portugal? *International Entrepreneurship and Management Journal*, 19, 781–803. https://doi.org/10.1007/s11365-023-00839-4.

NBR NM (2021). *Citizens Increasingly Enjoy the Benefits of Mobile Applications and E-Commerce*. NBR NM. https://www.nbrm.mk/ns-newsarticle-soopstenie-01072021-2-en.nspx.

NBR NM (2022). *Annual Information on Payments in the Republic of North Macedonia in 2022*. NBR NM. www.nbrm.mk/content/Platni%20sistemi/Презентација-Годишна-информација-за-плаќањата-2022г.pdf.

Neuman, D. (2014). Qualitative research in educational communications and technology: A brief introduction to principles and procedures. *Journal of Computing in Higher Education*, 26, 69–86. https://doi.org/10.1007/s12528-014-9078-x.

Normalini, M.K., & Ramayah, T. (2012). Biometrics technologies implementation in Internet banking reduce security issues? *Procedia-Social and Behavioral Sciences*, 65, 364–369. https://doi.org/10.1016/j.sbspro.2012.11.135.

Odorović, A., McKain, G., Garvey, K., Schizas, E., Zhang, B.Z., Rowan, P., & Ziegler, T. (2020). FinTech innovation in the Western Balkans: Policy and regulatory implications and potential interventions. *Available at SSRN 3619214*: https://ssrn.com/abstract=3619214 or http://dx.doi.org/10.2139/ssrn.3619214.

O'Neil, R., & Burt, C. (2017). Mobile biometric applications. *Biometrics Research Group*, 2017. https://www.biometricupdate.com/wp-content/uploads/2017/03/special-report-mobile-biometric-applications.pdf.

Palamidovska-Sterjadovska, N., Prodanova, J., & Ciunova-Shuleska, A. (2024). Why do customers value m-banking apps? A stimulus-organism-response perspective. *Spanish Journal of Marketing-ESIC*. https://doi.org/10.1108/SJME-01-2023-0024.

Parmar, V., Sanghvi, H.A., Patel, R.H., & Pandya, A.S. (2022, April). A comprehensive study on passwordless authentication. In: *2022 International Conference on Sustainable Computing and Data Communication Systems (ICSCDS)*, 1266–1275. https://doi.org/10.1109/icscds53736.2022.9760934.

Pons, A.P., & Polak, P. (2008). Understanding user perspectives on biometric technology. *Communications of the ACM*, 51(9), 115–118. https://doi.org/10.1145/1378727.1389971.

PwC. (2021). *Payments 2025 & Beyond: Navigating the Payments Matrix*. PwC. https://www.pwc.com/gx/en/industries/financial-services/publications/financial-services-in-2025/payments-in-2025.html.

Reportlinker. (2024). *Global Biometrics Market Overview 2024–2028*. Reportlinker. https://www.reportlinker.com/market-report/Security-Systems/13091/Biometrics?term=biometrics%20data&matchtype=b&loc_interest=9014312&loc_physical=2807&hsa acc=7466570189&hsa_cam=15062579447&hsa_grp=129922497023&hsa_ad=600845830002&hsa_src=g&hsa_tgt=kwd-1513224321009&hsa_kw=biometrics%20data&hsa_mt=b&hsa_net=adwords&hsa_ver=3&gad_source=1&gclid=CjwKCAjw7NmzBhBLEiwAxrHQ-U3y9rUISylTqZXXPF-bqi-40VyQhF-l7iLg5f_vUmKiPQSyJhTjXBoCecAQAvD_BwE.

Sheena, S., & Sheena, M. (2014). A study of multimodal biometric system. *International Journal of Research in Engineering and Technology*, 3(15), 93–98.

Smith, M., & Urbas, G. (2021). *Technology Law: Australian and International Perspectives*. Cambridge University Press.

Soto-Beltrán, L.L., Robayo-Pinzón, O.J., & Rojas-Berrio, S.P. (2022). Effects of perceived risk on intention to use biometrics in financial products: Evidence from a developing country. *International Journal of Business Information Systems*, 39(2), 170–192. https://doi.org/10.1504/ijbis.2022.121432.

Statista. (2023). Does your organization plan to adopt, or use/continue to use passwordless authentication in the next 1–3 years? Statista. https://www.statista.com/statistics/1441237/future-adoption-of-passwordless-authentication-by-various-industries-worldwide/

Suganya, M., Sujatha, S., & Alex, M.E. (2012). Secure user authentication using biometrics in mobile banking. *The Business & Management Review*, 2(1), 226.

Svilar, A., & Zupančič, J. (2016). User experience with security elements in Internet and mobile banking. *Organizacija*, 49(4), 251–260. https://doi.org/10.1515/orga-2016-0022.

Tanwar, S., Tyagi, S., Kumar, N., & Obaidat, M.S. (2019). Ethical, legal, and social implications of biometric technologies. *Biometric-based Physical and Cybersecurity Systems*, 535–569. https://doi.org/10.1007/978-3-319-98734-7_21.

Tassabehji, R., & Kamala, M.A. (2009, December). Improving e-banking security with biometrics: Modelling user attitudes and acceptance. In: *2009 3rd International Conference on New Technologies, Mobility, and Security,* 1–6. https://doi.org/10.1109/ntms.2009.5384806.

Telo, J. (2023). Understanding security awareness among bank customers: A study using multiple regression analysis. *Sage Science Review of Educational Technology*, 6(1), 26–38.

Toledano, D.T., Pozo, R.F., Trapote, A.H., & Gomez, L.H. (2006). Usability evaluation of multi-modal biometric verification systems. *Interacting with Computers*, 18, 1101–1122. https://doi.org/10.1016/j.intcom.2006.01.004.

Venkatraman, S., & Delpachitra, I. (2008). Biometrics in banking security: A case study. *Information Management & Computer Security*, 16(4), 415–430. https://doi.org/10.1108/09685220810908813.

Vivekanandan, L., & Jayasena, S. (2012). Facilities offered by the banks and expectations of IT savvy banking customers. *Procedia-Social and Behavioral Sciences*, 40, 576–583. https://doi.org/10.1016/j.sbspro.2012.03.233.

Vukomanović, D. (2021). Development of migration management portfolio in Western Balkans region. *Forum za sigurnosne studije*, 4(4/5), 97–123.

Wang, J.S. (2021). Exploring biometric identification in FinTech applications based on the modified TAM. *Financial Innovation*, 7(1), 1–42. https://doi.org/10.1186/s40854-021-00260-2.

Part II:

Challenges, Comparisons, and the Future of Biometrics

6

Knowledge and Perception of Biometric Payments by the Polish Public

Justyna Ziobrowska-Sztuczka[*(0000-0002-9572-1438)] **and Dominika Łyś**[(0009-0007-9909-0622)]

University of Wroclaw

1. Introduction

Responding to the rapid development of technology and the changing needs of consumers, who value the speed, convenience, and security of their transactions, it can be seen how payment systems are evolving. Technologies like Apple Pay, Google Pay, and Samsung Pay allow users to pay using their smartphones. Their popularity is growing due to the convenience and security of their transactions. So-called digital wallets, or digital wallets, are also emerging, namely, PayPal, Venmo, Alipay, and WeChat Pay, which allow multiple forms of payment to be stored in one place and offer fast and secure transfers. QR code payments, SMS, or M-commerce have unique features and can be used in different situations, depending on the users' needs. In different countries, the popularity and availability of each form of payment varies. Mobile wallets accounted for approximately ½ of global e-commerce payment transactions in 2023, meaning that this online payment method has become the most common form of payment worldwide. The mobile wallet market is projected to grow at a CAGR of 14.9% between 2023–2027 (E-commerce payment methods market share 2027; 2024). Mobile wallets have become very popular in two countries. Alipay and PhonePe

*Corresponding author: justyna.ziobrowska@uwr.edu.pl

play significant roles in payments in China, India, respectively. However, Apple Pay technology is prevalent in the US, UK, and Canada than in the rest of the world. It has been estimated that North America Apple Pay and Google Pay users will double between 2020–2025 (De Best, 2024). However, the most popular form of mobile payment in Poland is BLIK (Institute of Finance, 2022).

It is essential to remember that mobile payments involve the processing and storage of sensitive personal and financial data, and with their growing popularity, attempted fraud and hacking attacks are increasing (Bakunova et al., 2019). So-called cybercriminals seek to exploit vulnerabilities in security systems. It is, therefore, crucial to use robust authentication methods such as two-factor authentication (2FA) or biometrics (to which the current publication will be devoted) and one-time passwords to prevent unauthorized transactions. This article aims to analyze the degree of acceptance and willingness of Poles to use solutions like biometric payments and identify possible barriers associated with their introduction.

We decided to study the adoption of biometric payments in Poland, as the technology is growing rapidly worldwide, and its potential could revolutionize the way transactions are made. Analyzing this phenomenon in the Polish context will allow us to assess how ready the market and consumers are for the adoption of modern, secure payment methods.

The authors formulated the following research questions: What is the attitude of the Polish public towards new forms of payments made using biometric techniques? What are the potential opportunities for large-scale implementation of such technologies? Can biometric payments replace traditional methods? To find the answers, a diagnostic survey, literature analysis and critique were used, and a fundamental statistical analysis was performed.

2. Literature Review

The term 'biometrics' is derived from the Greek words bios-'life' and metron-'measurement'. The science deals with the measurement and analysis of a person's unique biological characteristics for identification and authentication purposes (Jaiswal, 2011; Marucha-Jaworska, 2015; Lewandowski, 2017; Mendyk-Krajewska, 2019). Human biometrics began to be used in the second half of the 19th century. However, the breakthrough came a century later with the development of information technology and the ability to process data quickly. Typical biometric identifiers include fingerprints, facial features, iris, retina, voice, and gait (Bolle, 2008; Ruiz Marín et al., 2009; Holyst, 2011; De Souza, 2020). The definition of "biometric data" is contained in Article 4(12) of the General Data Protection Regulation (GDPR). They denote personal data that result from specific technical processing, concern physical, physiological, or behavioral characteristics of an individual, and enable or confirm the unambiguous identification of that person. Biometric systems typically involve capturing a digital image, extracting distinguishing characteristics, and comparing them with stored patterns (Jaiswal et al., 2011). Applications of biometrics go beyond identification and include complex security systems, weapons identification, and medical diagnostics (Ruiz Marín et al., 2009). This technology is becoming increasingly important due to the limitations of current identification systems and the growing threat of identity theft (De Souza, 2020). Every interaction in a digital environment provides data on what we've done in that environment, playing a role in personalization, destination marketing, digital reputation, social media, and graphic services. Specifically, the digital footprint we leave behind is as large as the number of people or entities we interact with online (Liébana-Cabanillas et al., 2022). A digital identity can be defined as all online information and data about a specific person. Biometric verification compares a person's characteristics with their previously recorded identity and is faster and more efficient than identification with large databases (Bertolín, 2007). According to researchers, biometrics is a more effective alternative to traditional methods, namely, passwords or identification cards (Bertolín, 2007; Jaiswal, 2011). To evaluate the effectiveness of biometric methods, three main measures commonly used are:

1. FAR: Determines the probability that the biometric system will positively verify a situation (grant access) where it should not.
2. FRR: Defines the probability that the biometric system will reject an access attempt by a person authorized to access it. Both these measures should be as low as possible for extensive volume samples.
3. FTE: False registration rate.

In practice, finding a balance between FAR and FRR is primarily essential. FRR measures how often the system erroneously rejects eligible users. A decrease in FAR is often associated with an increase in FRR and vice versa. An ideal biometric system seeks to minimize both metrics (Czyżewski). Different authors use varied labels for these indicators (Gomółka, 2016; Mendyk-Krajewska, 2019; Szybalski, 2016) such as:

1. FMR: False match rate.
2. FNMR: False mismatch rate.
3. EER: Sustainable error rate.

Biometrics is becoming increasingly popular as a payment authentication method, and technological developments and the growing demand for secure and convenient payment methods suggest that its importance will grow in the future (Kumar & Ryu, 2008). These systems use unique biological features like fingerprints, iris patterns, or facial features to authenticate the user (Yang et al., 2013; Mróz-Gorgoń et al., 2022). It first appeared in banking in 2004, when Colombia's Bancafe Bank provided about 400 biometric ATMs using a fingerprint. In 2004, ATMs using vascular biometrics began were available in Japan (e.g., Mizuho Bank, Japan Post Bank, Bank of Kyoto, Resona Bank, and Bank of Yokohama). Iris biometrics found its first application in banking in Jordan (Cairo Amman Bank) – for ATMs and bank branches, followed by online banking. In Poland, biometrics in banking began in 2010, when Bank Polskiej Spółdzielczości and Podkarpacki Bank Spółdzielczy applied fingerprint biometrics (Lewandowski, 2017). Today, banks continue to explore biometric solutions to address security breaches and transactional fraud, which are becoming increasingly common in traditional systems (Venkatraman, 2008). Table 1 presents a brief history of the use of biometrics as a payment method. Successful implementation of biometric authentication in banking requires considering technological, managerial, social, and ethical factors (Venkatraman, 2008). As the digitization of the banking sector continues, biometrics is expected to play a vital role in ensuring information security and improving customer service (Bakunova et al., 2019).

Biometric payments have become increasingly popular among contactless payments in recent years. Poland is a country characterized by the dynamic development of biometric technologies, and its market is considered to be one of the most modern markets in Europe, as Polish society mainly makes cashless payments to a large extent (Mróz-Gorgoń et al., 2022). The global market for contactless biometrics technology (e.g., FaceID, iris pattern analysis, hand geometry, fingerprint) is valued at USD 20.4 billion in 2023. According to CARG, the market is projected to grow at a rate of 20% between 2024–2032, ultimately

Table 1: History of using biometrics as a form of making payments

Time-scale Calculus	Applications of Biometrics by Country
Second half of the 19th century	The first attempts to use human biometric traits for identification.
2004 year	First applications of biometrics in the banking sector: Colombia's Bancafe Bank introduced some 400 ATMs using fingerprint identification. In Japan, ATMs using vascular biometrics were launched at banks such as Mizuho Bank, Japan Post Bank, Bank of Kyoto, Resona Bank and Bank of Yokohama.
2010 year	In Poland Bank Polskiej Spółdzielczości and Podkarpacki Bank Spółdzielczy applied fingerprint biometrics for the first time.

Source: Own elaboration based on Bolle, 2008; Marín et al., 2009; Holyst, 2011; Lewandowski, 2017; De Souza, 2020.

leading to the market likely to be valued at USD 100 billion in 2032 (Contactless Biometrics Technology Market Size, 2024–2032 Report (gminsights.com); 2024).

It is worth noting that developments in artificial intelligence (AI) and machine learning (ML) are significantly improving the performance of biometric systems. AI in biometrics can detect and counteract fraud attempts (mainly spoofing) by recognizing the subtle differences between a person's true and false biometric characteristics. Additionally, the combination of AI and ML makes it possible to create biometric systems that use multiple biometric traits simultaneously, increasing the security and accuracy of these multimodal systems (Global Biometrics Market Size, Share & Growth Report [2031]).

3. Research Methods

The study was designed to determine Polish consumers' degree of knowledge and preferences related to biometric payments. The article has two parts. The first presents the essence and definitional issues related to biometric techniques and identifies trends in consumer behavior conditioning the development of the methods analyzed. The second presents the results of the authors' research into Polish consumers' experiences of biometric payments. The first stage of the study used the method of literature analysis and critique, and the second stage used the method of diagnostic survey and statistical analysis. The quantitative survey was conducted between 27 June 2024–12 September 2024 on 377 respondents using the standardized CAWI (Computer Assisted Web Interview) method. The survey questionnaire, made available electronically, required respondents to have access to the Internet, meaning that not everyone had the chance to be in the survey sample.

Based on the knowledge available at the article's preparation stage and the literature review, the research hypothesis is H1: *Age plays a major role in the acceptance of a new form of payment using biometric techniques.* To test this hypothesis, the authors conducted a correlation analysis of the responses from this characteristic's point of view (respondents' age).

4. Findings

A total of 377 respondents participated in the survey, of whom 71.4% were women and 28.6% were men. The largest group was represented by those in the 18–26 age range, accounting for 47.0% of the total. The survey also included those under 18 (1.1%), and respondents aged 27–36 (25.2%), 37–50 (15.6%), and over 50 (11.1%). Those with a university degree accounted for the most significant proportion (49.6%). Moreover, 26% of respondents declared that they were in the process of studying. Nonetheless, 19.6% respondents declared they had secondary education, while 4% had vocational education and 0.8% had primary education. Additionally, 21.4% of participants live in rural areas, and 17.5% of all respondents come from towns with a population of up to 50,000; 13.8% respondents live in towns with a population of 50,000 to 150,000, and 14.9% of all respondents live in towns with a population of 150,000 to 500,000. The largest group, which is exactly 32.4% of respondents, live in large cities with a population of more than 500,000.

Of the respondents, 89.4% said they were familiar with biometric payments, but only 38.7% of those surveyed used biometric techniques for various purposes not necessarily exclusively related to payment transactions. Among those using biometric techniques, the most popular technique is fingerprint (91%). A relatively significant proportion, 65.5%, said they also used a facial detail recognition method. The least popular biometric methods include voice tone recognition (7.6%) and iris scanning (2.8%).

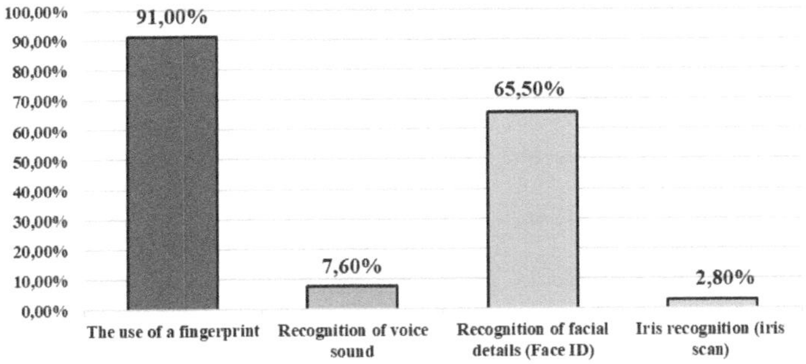

Figure 1: Biometric techniques used by Polish society.

Differences in the use of particular biometric techniques for making payments may be determined by the attitudes of Polish society towards specific methods, and also from the limited availability of devices enabling this type of transaction in some outlets. It should be noted that Poland is the first country in Europe to introduce a payment system based on the fusion of iris and facial biometrics. This innovation has a limited number of devices enabling such payments, with the result that the solution is currently only available in a few cities, namely Warsaw, Wrocław, Kraków, Poznań and Czeladź. (Biometric payments in Poland enter the next level, Strefa Biznesu, 2024.)

Another factor contributing to the low popularity of iris-based biometric payments in Poland is the fear of removing glasses or pulling out the contact lenses (31.0%). Additionally, 30.2% respondents believe that iris scanning may contribute to diseases such as photophobia (hypersensitivity to light), retinal damage from eye exposure to laser light, or dry eye syndrome. The comfort of use is essential in determining whether respondents would use specific biometric techniques to make payments; 22.3% respondents said they would feel completely comfortable using biometric payments, while only 10.3% rated the degree of use of such methods as entirely uncomfortable.

Based on the survey, Poles show a positive attitude towards biometric payments using a fingerprint (50.10%). In contrast, about half of the respondents express a negative attitude towards making payments through voice recognition (51.70%). Regarding biometric methods in the form of facial detail identification and iris scanning, respondents are ambivalent, indicating that they are hesitant and undecided in their assessment of these methods.

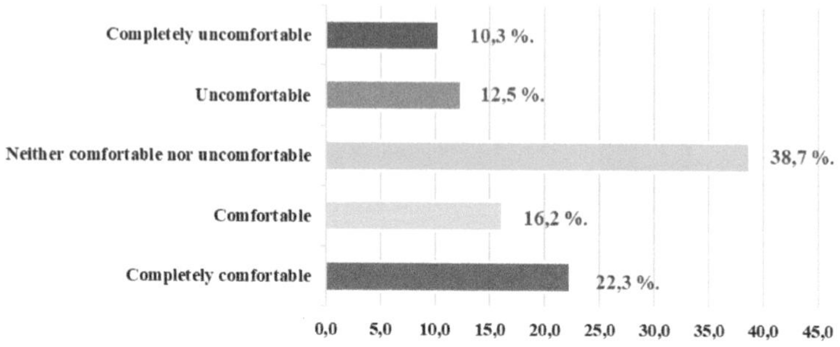

Figure 2: The comfort of use when making payments using biometric techniques according to respondents. (*Source:* Compiled from own research)

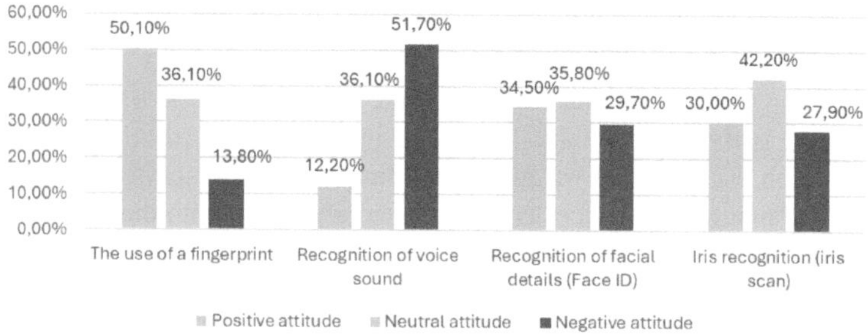

Figure 3: Polish public's attitude towards using specific biometric techniques as forms of payment. (*Source:* Compiled from own research)

Due to the potentially significant impact of age on the acceptance of modern forms of payment based on biometric technologies, it was decided to investigate how the different age groups of the Polish population perceive these innovative solutions and whether they are willing to accept biometric payments while abandoning traditional payment methods. Only those in the 27–36 and 37–50 age brackets express the belief that biometric payments will replace traditional transaction methods, becoming the dominant form of payment. The other age groups are more ambivalent on this question, with only minor percentage differences between responses. Among young people (under 18 and 18–26), there is a perception that currently available payment methods such as Blik, Apple Pay, Google Pay, and payment cards are better than modern biometric-based solutions. However, a significant proportion of the 18–26 age group, 35.6% exact, shows a positive attitude towards biometric payments, predicting that they could replace traditional forms of payment in the future. Similar attitudes can be observed among the over-50s. The younger generation was predicted to be most open to accepting biometric payment methods, while older people were expected to remain attached to existing solutions. Nevertheless, the survey results indicate that middle-aged people manifest the highest acceptance of biometric payments. At the same time, positions are mixed in the other age groups, and opinions remain divided.

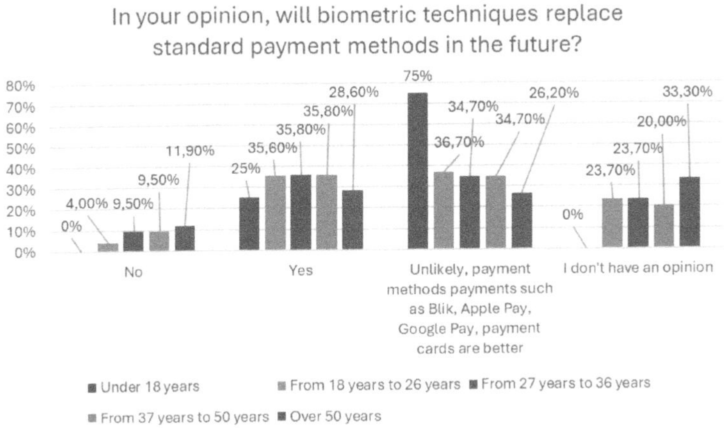

Figure 4: Polish public's attitude towards accepting biometric payments.

This attitude of the Polish public towards biometric payments can be linked to several concerns, particularly the fear of losing identity. The Polish public perceives the primary risk associated with using biometric techniques in payment processing to be that such systems may wrongly identify unauthorized persons as authorized, leading to unauthorized access to funds (62.9%). Additionally, respondents point to the risk of biometric data being intercepted by unauthorized individuals during transfer between devices, potentially leading to identity theft (57.6%). The belief that health problems such as conjunctivitis or dry eye syndrome could prevent payments from being made using an iris scan is also a significant concern (46.2%). In times of digital revolution, the main concern is the development of technology, especially the development of AI. The Polish public is also concerned about the possibility of technical modifications and voice manipulation, which

Table 2: Polish public's concerns about biometric payments

The Potential Risks Associated with the use of Biometrics for Payments	% of Respondents in the Age					% of the Total Number of Respondents
	Under 18 age	18–26	27–36	37–50	Over 50 age	
Systems may mistakenly identify unauthorized persons as authorized, leading to unauthorized access	0.0%	72.3%	66.3%	50.8%	38.1%	62.9%
Invasion of privacy	0.0%	45.8%	36.8%	40.7%	38.1%	41.4%
When transferring biometric data between devices, there is a risk of interception by unauthorized persons (identity theft)	50.0%	68.4%	47.4%	49.2%	47.6%	57.6%
Implementation of biometric technologies can be costly	0%	18.6%	26.3%	20.3%	23.8%	21.2%
In order to make payments using the iris of the eye, it will be necessary to take out contact lenses or remove glasses	25.0%	36.7%	17.9%	28.8%	38.1%	30.8%
Development of mobile applications designed to scan the iris of the eye to make payments using biometric methods during online purchases, increasing the risk of identity impersonation	0.0%%	41.8%	30.5%	20.3%	31.0%	34.0%
Frequent failure rate of the newly developed system	0.0	35.6%	25.3%	13.6%	23.8%	27.9%
Inability to make payment in case of health problems (e.g., conjunctivitis, dry eye syndrome)	25.0%	54.8%	33.7%	40.7%	47.6%	46.2%

Source: Own elaboration based on the diagnostic survey conducted.

could facilitate impersonation. Of particular concern is the prospect of introducing mobile apps that allow iris scanning to make payments when shopping online, which respondents believe will further increase the risk of impersonation (34.0%). Respondents even reported concern in terms of an increased number of murders to steal money using biometric data, and concerns about the frequent failure rate of the newly developed system (27.9%).

Despite the concerns, it was decided to investigate whether the Polish public would consider using biometric payments in the future in the event of adequate security and widespread availability. The survey found that mainly middle-aged people, i.e., those aged 27–36 (42.1%) and 37–50 (57.6%), are willing to use biometric payments provided they are secure. Importantly, contrary to predictions, older people (over 50) do not show an attachment to traditional forms of payment (only 4.8% of this group) nor a negative attitude towards biometric payments. In addition, 47.6% of those aged 50 and over and 50% of those under 18 said they were likely to use biometric methods to make payments in the future, provided they were secure and widely available. However, half of the under-18 age bracket said they were attached to traditional forms of payment, i.e., cash.

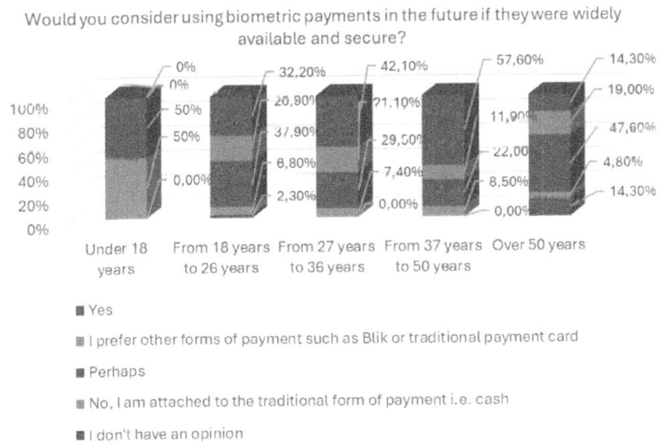

Figure 5: The willingness of the Polish public to use biometric techniques to make payments, provided they are adequately secure and widely available.

The Polish public is therefore considering using biometric payments in the future and sees specific advantages in them. Poles believe that making payments using biometric techniques is better than existing methods because they do not need to carry a smartphone or wallet to make a transaction (69.6%). Additionally, respondents value significantly better transaction speed (53.1%) and convenience and hygiene (25.5%) in this method. These factors are of particular importance in the context of global crises such as the pandemic caused by the SARS-CoV-2 virus. The Polish public also expresses a desire to use biometric payments because they are not associated with the need to remember PINs or passwords to confirm transactions (0.8%) and biometric techniques show a high degree of adaptation to different user groups, mainly the elderly and people with disabilities (11.1%). Respondents also see a crucial advantage in making payments through iris scanning, as Poles perceive the security of this method because the iris of the eye is unique and special to each individual (22%) and the the iris does not change throughout a person's life, eliminating the risk of identity impersonation (23.6%).

5. Discussion

As Szybalski (2016) notes, payment authentication methods using biometrics are more convenient and faster than traditional solutions in the form of passwords and PINs. The authors' survey results also confirm this, as 24.1% respondents declared that the speed of the transaction process is the most essential feature in a payment system based on biometric techniques. In turn, 29.4% respondents believe that biometric payments are better than traditional solutions due to the convenience and intuitiveness of using such a system.

Biometric techniques make it possible to quickly verify the identity of a specific person, as this is not linked to the need to remember passwords or PINs, as their role in this case, e.g., can be performed by voice. Furthermore, verification using biometrics is, on average, 20 seconds faster than traditional methods (Szybalski, 2016). Thus, payment authentication methods using biometric techniques can be faster and more secure than classic payment methods. For example, one of the descriptors of electronic voice analysis is the assessment of the speaker's emotional state (Slot, 2010). A particular person's pronunciation technique is also considered in this payment authentication method, so the accent and the rate at which individual words are spoken are important (Grzybowska, 2012). On the other hand, identifying and confirming the identity of a specific person using hand geometry uses the identification of individual characteristics such as the width of the metacarpal and the length and thickness of the fingers (Gomółka et al., 2016). The potential risk of misidentification in this method is ruled out by the fact that these devices are sometimes equipped with technology to measure the temperature, stability, and conductivity of the skin of the hand. It is important to emphasize that this method is very effective, as changes associated with aging and weight gain do not present a barrier or potential risk of misidentification of a person (Grzybowska, 2012). Facial detail recognition is another method of payment authentication. Identity is verified based on a mathematical model of the face built (created during the verification phase), which is then compared with the model recorded during the registration phase (Grzybowska, 2012). During identification, the person's individual characteristics are considered, like distance between the eyes and the edges of the eyes and the nose, and the eyes and the mouth. The shape of the facial contours visible in the profile and from the front, and the width of the face, are also accounted for (Ambroziak, 2002). An important aspect, especially for women, of the effectiveness of this method is that make-up or hairstyles done differently do not prevent correct identification (Grzybowska, 2012). Even though the method of authenticating payments by recognizing facial details is highly secure, only 34.5% of the authors' study respondents showed a positive attitude towards this biometric technique. Therefore, such a method for completing transactions in Poland has minimal prospect of adaptation.

However, an iris scan is considered the most effective method of proper verification, allowing access by authorized persons (Grzybowska, 2012; Mlaskawa, 2015). This is possible due to the great diversity and uniqueness of the iris texture. Furthermore, the biological variability of iris texture over the years is impossible from a practical point of view, and the probability of it changing is almost zero (Strąk et al., 2014). It should be emphasized that fingerprinting is still the most popular identification method. This is confirmed by the results of the authors' study as 91% respondents using biometric techniques declared that they had used this method. A fingerprint scanner makes it possible to compare a user's fingerprint with specific data stored in a database in a fraction of a second. This device considers the tension of the epidermis and temperature differences between the convex and concave parts of the fingerprint, effectively preventing fraud (Grzybowska, 2012).

Numerous studies confirm that users are willing to use biometric methods widely, provided that this increases their security. In 2014

(10 years ago), a study conducted by TNS Poland on behalf of Nuance Communications showed that nearly half of the respondents (49%) were willing to use biometric methods instead of passwords and PIN codes. Half of the respondents emphasized the lack of user influence on the security of the data they provided (Balawender, 2015). Contrastingly, based on the authors' survey, 33.4% respondents believe that biometric payment processing is as secure as other methods. Furthermore, only 22.5% respondents believe that payments using biometric techniques are more secure than traditional forms of payment. The most crucial security standards in the context of payments using biometric techniques are declared by respondents to be advanced data encryption methods that provide double security (72.4%) and double authentication, i.e., a created individual code that is assigned to a person (59.9%), and fast operator response to all kinds of incidents and payment system failures (43.5%). However, 66.6% respondents believe introducing appropriate legal regulations and security standards could reduce public concerns about payments using biometric techniques[1].

6. Conclusions

There are grounds to reject H1: *Age plays a significant role in accepting a new form of payment using biometric techniques,* as confirmed by the author's research and the literature. Predominantly, 18–26-year-olds use multiple biometric techniques, mainly fingerprint and facial detail recognition. The attitude of the Polish public towards biometric techniques in the form of fingerprint and voice sound recognition varies. Poles are positively disposed towards biometrics using fingerprints, while the identification method based on voice tone recognition raises concerns and provokes adverse reactions among the public. However, those most open to accepting and using biometric payments in the future are those aged 27–36 and 27–50. The Polish public perceives biometric payment processing as having advantages, namely, not having to carry a smartphone or wallet to make a transaction, the convenience, and

[1] The authors conducted the survey on a smaller research sample compared to the TNS Poland survey. The representative sample was 377 and there were no age restrictions, which means that the research sample also included people under the age of 18. The survey questionnaire was made available electronically, which means that the survey only includes people with Internet access.

intuitiveness of using a biometric payment system and a much faster payment process, and above all, the adaptability to different user groups mainly for the elderly and people with disabilities. It should be emphasized that two concerns about the processing of payments using biometrics are prevalent in the Polish population regardless of age, namely, that the systems may wrongly identify unauthorized persons as authorized, leading to unauthorized access to funds, and they point to the risk of biometric data being intercepted by unauthorized persons during transfer between devices. Furthermore, Poles are concerned that health problems such as conjunctivitis or dry eye syndrome will prevent them from making payments. Additionally, a significant source of their concern is the potential privacy risks associated with using these technologies.

Various factors, such as education, security, privacy, and user experience, are vital to increasing people's openness to biometric payments. To increase trust, companies must ensure that biometric data is stored securely, like using encryption or storing data locally on the device rather than in the cloud. Increasing the number of outlets supporting biometric payments may lead users to use the technology more frequently. We believe that "investment in education and awareness campaigns about biometric payment security" could be a way forward. Institutions should conduct educational campaigns explaining the advantages and security of biometric payments to build customer confidence in this new technology. It is worthwhile to introduce biometric payments in selected market segments or locations to minimize risks and monitor consumer reactions before widespread adoption.

References

Ambroziak, R. (2002). Biometryczne systemy identyfikacji. *Acta Universitatis Lodziensis, Folia Oeconomica*, vol. 157.

Bakunova, T., Trofimova, E., & Lapteva, E. (2019). Biometrics as a method of information security in the banking sector digitalization. In: *Proceedings of the International Scientific and Practical Conference on Digital Economy (ISCDE)*. https://doi.org/10.2991/iscde-19.2019.50.

Balawender, D. (2015). Nowy raport TNS pokazuje, że Polacy mają dość haseł dostępu, numerów PIN, tokenów. *Czas na hasło głosowe*. Pobrane z: http://banking-magazine.pl/2015/01/15/nowy-raport-tns-pokazuje-ze-polacy-maja-dosc-hasel-dostepu-numerow-pin-tokenow-czas-na-haslo-glosowe/ (2024.09.14).

Bertolín, J.A., & Bertolín, M.A. (2007). Análisis en torno a la tecnología biométrica para los sistemas electrónicos de identificación y autenticación. *Revista española de electrónica.* ISSN 0482-6396, pp. 52–67.

Bodowski, M. (2024). Wkrótce w Polsce zapłacisz z wykorzystaniem biometrii tęczówki oka i twarzy. *Rusza program pilotażowy. To najbezpieczniejsze rozwiązanie.* https://strefabiznesu.pl/wkrotce-w-polsce-zaplacisz-z-wykorzystaniem-biometrii-teczowki-oka-i-twarzy-rusza-program-pilotazowy-to-najbezpieczniejsze/ar/c3-18583625. Access date: 18.09.2024.

Bolle, R.M., Connell, J.H., Pankanti, S., Ratha, N.K., & Senior, A.W. (2008). *Biometria.* Wydawnictwa Naukowo – Techniczne, Warszawa.

Contactless Biometrics Technology Market Size, 20242032 Report. https://www.gminsights.com/industry-analysis/contactless-biometrics-technology-market. Access date: 17.07.2024.

Czyżewski, A., & Hoffmann, P. (2016). Przegląd technologii biometrycznych, prezentacja PowerPoint. Pobrano z: https://multimed.org/student/pdio/pdio10_biometria.pdf. data dostępu: 17.07.2024.

De Best, R. (2023). Most Used Payment Methods in E-commerce Worldwide in 2023, with a Forecast for 2027. https://www.statista.com/statistics/1111233/payment-method-usage-transaction-volume-share-worldwide/. Access date:17.07.2024.

De Best R.(2023). Most Used Payment Methods in E-commerce Worldwide in 2023, by Region. https://www.statista.com/statistics/348004/payment-method-usage-worldwide/. Access date: 17.07.2024.

Global Biometrics Market Size, Share, & Growth Report, 2031. https://www.kingsresearch.com/biometrics-market-253. Access date: 18.07.2024.

Gomółka, Z., Twaróg, B., & Żesławska, E. (2016). Identyfikacja tożsamości z wykorzystaniem analizy geometrii dłoni. *Journal of Education, Technology and Computer Science*, 18(4). https://doi.org/10.15584/eti.2016.4.56.

Grzybowska, A. (2012). Nowoczesne technologie w rozwoju usług sektora bankowego. Zeszyty Naukowe Uniwersytetu Szczecińskiego. Ekonomiczne Problemy Usług, No. 96.

Hołyst, B., & Pomykała, J. (2011). Biometria w systemach uwierzytelniania. *Biuletyn WAT*, 60(4), 418–419.

Instytut Finansów. (2022). Płatności gotówkowe i bezgotówkowe. Analiza stanów i perspektyw. Raport 2/2022. Warszawa.

Jaiswal, S., Bhadauria, S., & Jadon, D.S. (2011). Biometric: Case study. *Journal of Global Research in Computer Sciences*, 2(10).

Kumar, D., & Ryu, Y. (2008). A brief introduction of biometrics and fingerprint payment technology. In: *Second International Conference on Future Generation Communication and Networking Symposia.* https://doi.org/10.1109/FGCNS.2008.11.

Lewandowski, R. (2017). Biometria – nowe zastosowania. *Przegląd Bezpieczeństwa Wewnętrznego*, 9(17), 152–166.

Liébana-Cabanillas, F., Muñoz-Leiva, F., Molinillo, S., & Higueras-Castillo, E. (2022). Do biometric payment systems work during the COVID-19 pandemic? Insights from the Spanish users' viewpoint. *Financial Innovation*. https://doi.org/10.1186/s40854-021-00328-z.

Marin, L., Ruiz, S., & Rubio, A. (2009). The role of identity salience in the effects of corporate social responsibility on consumer behavior. *J Bus Ethics*, 84, 65–78. https://doi.org/10.1007/s10551-008-9673-8

Marucha-Jaworska, M. (2015). Podpisy elektroniczne, biometria, identyfikacja elektroniczna. *Elektroniczny obrót prawny w społeczeństwie cyfrowym*. Wolters Kluwer, Warszawa.

Mendyk-Krajewska, T. (2019). *Biometryczne metody sprawdzania tożsamości w nowych zastosowaniach*. Pobrano z: https://rocznikikae.sgh.waw.pl/p/roczniki_kae_z54_03.pdf.

Mlaskawa, J. (2015). *Biometria w bankowości - szanse i zagrożenia Banku przyszłości*. Zeszyty Naukowe Polskiego Towarzystwa Ekonomicznego w Zielonej Górze, No 3.

Mróz-Gorgoń, B., Wodo, W., Andrych, A., Caban-Piaskowska, K., & Kozyra, C. (2022). Biometrics innovation and payment sector perception. *Sustainability*, https://doi.org/10.3390/su14159424.

Rozporządzenie Parlamentu Europejskiego i Rady (UE) 2016/679 z dnia 27 kwietnia 2016 r. w sprawie ochrony osób fizycznych w związku z przetwarzaniem danych osobowych i w sprawie swobodnego przepływu takich danych oraz uchylenia dyrektywy 95/46/WE.

Strąk, M., & Płaczek, J. (2014). *Bezpieczeństwo w bankowości elektronicznej dla klientów indywidualnych na podstawie banku Millennium S.A.* Zeszyty Naukowe Wyższej Szkoły Informatyki, Zarządzania i Administracji w Warszawie, No. 2 (27).

Souza, M.A.A. (2020). *A Biometria e suas Aplicações*. Revista Brasileira de Ciências Policiais.

Szybalski, M. (2016). *Biometria- nowoczesne metody uwierzytelniania*. White Paper. https://forumcallcenter.pl/wp-content/uploads/2017/11/ICT-Biometria-Whitepaper-PL.pdf.

Venkatraman, S., & Delpachitra, I. (2008). Biometrics in banking security: A case study. *Information Management & Computer Security*, 16(4), 415–430. https://doi.org/10.1108/09685220810908813.

Yang, W., Hu, J., Yang, J., Wang, S., & Shu, L. (2013). Biometrics for securing mobile payments: Benefits, challenges, and solutions. In: *International Congress on Image and Signal Processing*. https://doi.org/10.1109/CISP.2013.6743950.

7

Fingerprint vs. Voice

Francisco Liébana-Cabanillas[1*(0000-0002-3255-0651)], **Juan F. Prados-Castillo**[1(0000-0003-3588-5534)], **Francisco Muñoz-Leiva**[1(0000-0002-4996-7525)] **and Aijaz A. Shaikh**[2(0000-0001-5389-4384)]

[1] University of Granada, Spain

[2] The Institute of Information and Computational Technologies, Almaty, Kazakhstan

1. Introduction

Technological advances in biometric verification have revolutionized the manner in which financial transactions are conducted (Bhat et al., 2024). Biometric payment systems that use unique human physical characteristics to authenticate users are rapidly gaining ground in the financial sector owing to their increased security and convenience (Ahmed et al., 2021). The most promising biometric modalities for payment are fingerprinting and voice recognition (Kambampati et al., 2024).

Several terms have been used interchangeably to represent biometric payment systems, namely, biometric verification payment systems (Wadood et al., 2017), biometric authentication (Kumar, 2024), biometric identification (Wang, 2021), biometric recognition (Lee & Pan, 2023), and biometric contactless (Hossain and Muhammad, 2017). For this study, we have used the term "biometric payment systems" that allow users to conduct transactions using their unique physical characteristics like fingerprints, facial recognition, and iris scans. This technology offers a combination of convenience and security by eliminating the need for traditional passwords or PINs and reducing the risk of fraud. According to a recent report by Juniper Research (2021), biometric

*Corresponding author: franlieb@ugr.es

payments are expected to significantly exceed 1.4 billion transactions by 2025, from the 671 million recorded in 2020. This growth is largely due to the adoption and use of biometric-enabled mobile devices and increased user confidence in these technologies.

Fingerprint payments have recently gained popularity, particularly through the integration of sensors and bank cards. They allow the authentication of the payment cardholder without additional effort, unlocking the potential for contactless payments for high-value transactions (Ogbanufe & Kim, 2018). However, voice-recognition payment systems do not require physical contact and can be used remotely (Kathuria et al., 2020).

Both technologies have unique advantages and challenges. Although fingerprints are considered highly accurate and difficult biometrics to forge, voice recognition may be susceptible to environmental variations or changes in a user's voice (Kumar et al., 2021). However, advances in artificial intelligence (AI) are constantly improving the accuracy and security of both types of biometric systems (Kumari, 2024).

Simultaneously, European regulations such as the Revised Payment Services Directive (PSD2) and under consideration, PSD3, have promoted and motivated the implementation of biometric payment systems across member states by requiring Strong Customer Authentication (SCA) mechanisms. PSD2, implemented in May 2018 after the Cambridge Analytica Scandal that compromised over 87 million Facebook account users, introduced multifactor authentication to enhance the security of electronic transactions, and biometric methods have proven to be an efficient solution for these requirements. PSD3 seeks to address new challenges in the security of digital and mobile payments, including biometric-based transactions, thus ensuring an up-to-date regulatory framework that protects consumers while fostering technological innovation (European Commission, 2023).

Despite the growing interest in these payment methods, the academic literature lacks information on user perceptions and preferences for different biometric payment systems. This study seeks to fill this gap through an online survey that directly compares fingerprint and voice verification payment systems from the user's perspective. To address this research gap, this study pursues the following objectives.

RO1: Contribute to deepening the understanding of consumer behavior related to biometric payment systems and identify the determinants and barriers affecting their use based on Zarco et al.'s (2024) proposals.

RO2: Develop a conceptual framework to better understand the antecedents of the intention to use biometric payment systems based on UTAUT2, including the major variables Perceived Risk and user trust.

RO3: Analyze the differences in the intention to use fingerprint and voice payment systems.

This paper is organized as follows. Section 2 defines the key terms used, summarizes the importance of biometric payment systems, and describes the theoretical framework. Section 3 describes the formal methodology for data collection, survey design, and metrics used in defining the problem. Finally, Section 4 present the results of the analysis and discussion, respectively, along with the implications, limitations, and future research directions.

2. Literature Review

2.1. Biometric Payment Systems

Biometric payment systems have transformed the manner in which financial and nonfinancial transactions are verified and conducted. These technologies enable authentication through the unique characteristics of the individual, namely, fingerprints, facial recognition, iris, and voice, facilitating the fast and accurate user identification. A study by Allied Market Research (2022) projects that the global biometric payments market will reach USD 24.6 billion by 2026, driven by the growing use of mobile devices equipped with biometric technologies and the demand for greater security in financial transactions.

Biometric payment systems are divided into two broad categories: physiological and behavioral (Hamid et al., 2024). The former uses the physical characteristics of the individual, whereas the latter bases their authentication on behavioral characteristics, namely, walking or talking. Fingerprinting is one of the most widely used forms of biometrics in payment systems owing to the uniqueness of each individual's fingerprint and its easy integration into mobile devices and payment terminals. According to Statista (2024), approximately 80% of smartphones in the market today have integrated fingerprint technology, facilitating their adoption for mobile payments. This allows users to authorize payments quickly and securely by scanning their fingerprints on a device, thereby eliminating the need for passwords or PINs. Its accuracy and low margin of error have been widely validated, increasing the confidence of both consumers and financial service providers (Kim et al., 2022).

However, voice recognition is an emerging verification technology allowing users to authenticate digital payments through behavioral biometrics. It analyzes the unique characteristics of a person's voice, like pitch, frequency, and rhythm, to confirm their identity (Gómez-Barrero et al., 2017). One main advantage of this verification system is its application in screenless devices, namely, smart speakers and virtual assistants, allowing for completely hands-free payments (Guan et al., 2023). The adoption and use of these systems is expected to grow at a compound annual growth rate of 20.5% between 2021 and 2026, driven by the increased use of virtual assistants like Amazon Alexa and Google Assistant (Allied Market Research, 2022).

Biometric payments have numerous advantages over other payment systems (Faster Capital, 2024): (1) They provide a higher level of protection than conventional methods. As biometric features are inherently personal and unidentifiable, the possibility of fraudsters gaining access to accounts is significantly reduced. This has led to a marked decrease in the risk of identity theft and other fraudulent activities. (2) They significantly increase the transaction speed. Instead of searching for and handling cards or cash, users can complete their payments with a simple gesture like a tap or scan. This streamlined process reduces waiting times and substantially improves customer satisfaction during shopping. (3) They eliminate the need to constantly carry physical payments. By linking biometric features directly to bank accounts or credit cards, users can seamlessly perform transactions without relying on tangible objects, greatly simplifying their daily routines. (4) Biometric payments offer an inclusive solution for those who encounter difficulties using traditional methods. People with certain disabilities who might have trouble handling cash or cards find biometrics a more accessible and user-friendly alternative, promoting financial independence. (5) The horizon for biometric payments is promising and constantly expanding. Significant improvements are anticipated with continued advancement and increasing adoption of technology. This could include even greater authentication accuracy and the possibility of using multiple biometric identifiers for the same account, thus offering unprecedented flexibility and security in the world of payments.

2.2. UTAUT2, Perceived Risk Theory, and Trust Theory

The UTAUT2 model (Venkatesh et al., 2012), an extension of the original UTAUT, is widely used to analyze technology acceptance in the consumer context. It incorporates seven key constructs that influence technology adoption: expected performance, expected effort, social influence, enabling conditions, hedonic motivation, price, and habits. These factors play a crucial role in understanding how and why users adopt or reject biometrics technologies. For example, expected performance reflects perceptions of the efficiency and security of biometric payments, whereas expected effort considers ease of these systems. Habit and enabling conditions like technological infrastructure also determine their adoption (Venkatesh et al., 2012).

The UTAUT2 model has been extensively analyzed in the scientific literature and is considered by many authors as the most appropriate approach to study technology adoption (Higueras-Castillo et al., 2023). It has been used in various studies on payment systems (Al-Okaily et al., 2023). Table 1 presents a description of the variables in the UTAUT2 model.

Additionally, the inclusion of Risk and Trust was proposed because they are closely related to the adoption of financial technologies (Belanche et al., 2022; Migliore et al., 2022; Zarco et al., 2024; Blanco-Oliver et al., 2024; Ling et al., 2024). However, risk theory in the field of technology and payment systems refers to users' perceptions of potential losses or negative consequences when using new technologies (Pavlou, 2003). According to the literature, perceived risk is closely related to the adoption of biometric payment systems, as users tend to be more cautious due to concerns about privacy, the security of their biometric data, and the potential misuse of information (Featherman & Pavlou, 2003; Chukwuere et al., 2021; Hu et al., 2023).

Trust is another fundamental concept in the adoption of biometric payment systems. This refers to the user's belief that the technology is secure, reliable, and will work as expected. Trust theory argues that, in the context of digital payments, trust in the service provider, technology, and regulatory institutions significantly influences users' willingness to adopt new technologies (McKnight et al., 2002). Previous studies have shown that trust can mitigate the negative effect of perceived risk, increasing the intention to use innovative systems, namely, biometric payments (Gefen et al., 2003; Shiau et al., 2023).

Table 1: Variables of the UTAUT2 model

Construct	Definition	Key Features
Effort Expectancy	It refers to the degree of ease associated with using a particular system or technology. It reflects users' perceptions of how effortlessly they will adopt and operate technology.	Ease of use; User-friendly interface; Relevance in early adoption
Performance Expectancy	It refers to the degree to which an individual believes that using a particular system or technology will help them achieve gains in job performance or personal productivity.	Perceived usefulness; Task efficiency; Critical predictor of adoption
Hedonic Motivation	It refers to the degree of pleasure or enjoyment derived from using a particular system or technology.	Enjoyment and fun; Intrinsic motivation; Importance in consumer contexts
Social Influence	Social influence refers to the degree to which an individual perceives that important others (e.g., friends, family, colleagues, or societal norms) believe they should use a particular system or technology.	Subjective norms; Peer and social norms; Cultural sensitivity
Habit	It refers to the extent to which people tend to perform behaviors automatically because of learning and repetition.	Past behavior; Role in retention
Facilitating conditions	It refers to the degree to which an individual believes that organizational and technical infrastructure exists to support the use of the system or technology.	Resources and support; System compatibility; Influence on use behavior

Note: In our theoretical approach, we have not included the price variable because in the country where the research has been carried out, it is a new technology that has a direct cost for users.
Source: Venkatesh et al. (2012).

2.3. Comparative Analysis of Fingerprint and Voice Payment Systems

Several studies in the field of payment technology adoption have shown that specific features influence user perceptions in terms of trust, perceived effort, and overall acceptance (Ogbanufe & Kim, 2018; Shiau et al., 2023). Biometric payment systems, like fingerprint and voice recognition, have unique properties that affect not only perceived ease of use but also perceptions of security and convenience. These technological differences may modify the impact of key variables in the UTAUT2 model on usage intention, as users adjust their attitudes and behaviors according to the type of system proposed to them (Kim et al., 2022; Hu et al., 2023).

Voice recognition enables contactless interactions, which may be perceived as more convenient in certain situations, especially in environments where hygiene or minimal contact is a priority (Kambampati et al., 2024). Conversely, fingerprint technology, widely integrated into mobile devices and other systems, is seen as a trusted and familiar form of authentication that can facilitate its acceptance owing to the familiarity and low complexity (Gomez-Barrero et al., 2017). Thus, the type of biometric system modifies the perception of factors like expected effort or perceived trust, and also alters the intensity with which these factors impact the intention to use them.

Although several studies have proposed a comparative analysis between mobile payment systems, namely, SMS, NFC, and QR (Liébana-Cabanillas et al., 2017; Ramos de Luna et al., 2019), our research establishes a novelty by comparing two biometric payment systems. This approach suggests that the proposed biometric payment system moderates key relationships between model constructs, like performance expectancy, expected effort, social influence, perceived trust, and intention to use. Consequently, users are expected to experience variations in their intention to adopt these systems, depending on their perceptions of the specific type of biometrics used. Figure 1 and Table 2 present the model and the hypotheses to be tested.

Table 2: Hypothesis to be tested

No.	Hypothesis	Source
H1 (+)	Effort expectancy has a positive effect on the intention to use BPS	Venkatesh et al. (2003); Taiwo and Downe (2013); Williams et al. (2015)
H2 (+)	Performance expectancy has a positive effect on the intention to use BPS	Venkatesh et al. (2003); Nguyen and Le (2022); Yadav and Sharma (2023)
H3 (+)	Hedonic motivation has a positive effect on the intention to use BPS	Venkatesh et al. (2003); Liébana-Cabanillas et al. (2021); Seng and Hee (2021); Liébana-Cabanillas et al. (2023)
H4 (+)	Social influence has a positive effect on the intention to use BPS	Venkatesh et al. (2003); Molinillo and Higueras-Castillo (2022); Liébana-Cabanillas et al. (2023); Slade et al. (2015)
H5 (+)	Habit has a positive effect on the intention to use BPS	Venkatesh et al. (2003); Molinillo and Higueras-Castillo (2022); Liébana-Cabanillas et al. (2023); Slade et al. (2015)
H6 (+)	Facilitating conditions has a positive effect on the intention to use BPS	Venkatesh et al. (2003); Lisana (2021); Madan and Yadav (2017); Oliveira et al. (2016)
H7 (-)	Perceived risk has a negative effect on the intention to use BPS	Soto-Beltrán et al. (2022); Bhat et al. (2024)
H8 (+)	Perceived trust has a positive effect on the intention to use BPS	Liébana-Cabanillas et al. (2022); Bhat et al. (2024)
H9	Users' behavior towards the proposed BPS will differ from one system to another	

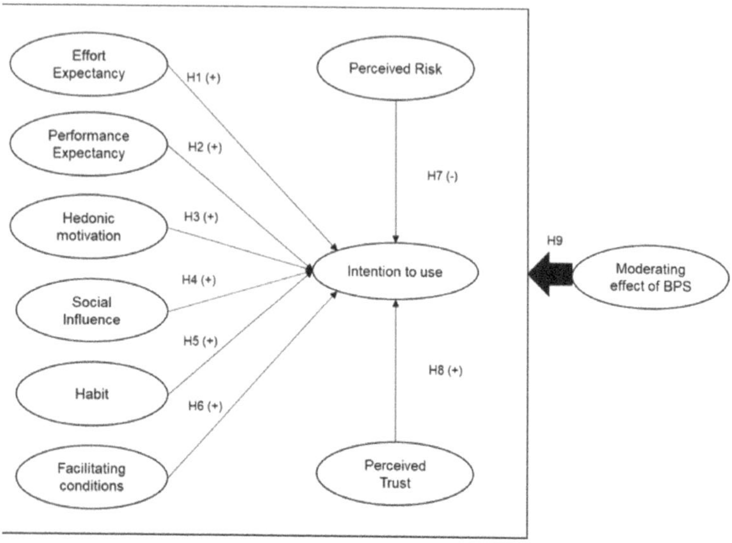

Figure 1: Proposed model.

3. Research Methodology

3.1. Measurements

The questionnaire included a series of statements that participants were asked to evaluate on a 7-point Likert scale, where 1 corresponded to "strongly disagree" and 7 to "strongly agree". This allowed us to examine the factors identified in the literature review. The measurement scales were adjusted based on previous studies. All variables in the UTAUT2 model were assessed based on the principles of Venkatesh et al. (2012). To measure perceived risk, an adaptation of the Singh et al. (2021) scale was used, along with the scale adapted from Liébana-Cabanillas et al. (2020). As we used a Spanish-translated version of the questionnaire, a professional native translator translated the final version into English to ensure consistency between the two.

3.2. Sample and Data Collection

Data were collected through a self-administered online survey between June 2024 and September 2024. After cleaning the data, the final sample comprised 224 potential biometric payment users. Convenience sampling (non-probability) was used, with a link to the survey distributed via social media and email lists.

First, an initial review (first phase) of both the methodology to be used and the scales extracted from the scientific literature was conducted by a panel of 10 experts. Based on personal interviews, this review led us to modify some of the proposed scales and adapt others to the main objective. Following this review by experts, we divided the questionnaire into three thematic blocks: control questions, questions about the object of the research, and socio-demographic data. The questionnaire related to each of the two payment systems analyzed was presented randomly.

An explanatory video was generated for each payment system analyzed to ensure that participants understood and answered the questionnaire correctly. While presenting the survey to the participants, they were informed that the video would contain a code that they had to enter for some of the questions in the questionnaire. Only responses that correctly indicated the code in question were used. According to Wells (2014) and Liébana-Cabanillas et al. (2018), consciously or unconsciously processed information stimulates recall, increasing the likelihood that participants will remember the messages they have shown, thus improving the reliability of the results.

To minimize the dropout rate, we included information about the research's purpose in the survey and a statement guaranteeing respondents' anonymity. We also offered small incentives, like mugs and umbrellas with the university logo, to randomly selected participants at the end. To reduce the incidence of missing values, participants were asked to answer all questions/statements before moving to the next page/end of the survey (otherwise, a notification was displayed). Regarding the characteristics of the sample, 58% were female, 44.6% were between 18 and 24 years old, with an income level of less than 1,500 euros (42.85%), and a university level of education (72.76%).

3.3. Data Analysis and Results

We employed partial least squares structural equation Modeling (PLS-SEM) to analyze the responses obtained. This cause-predictive approach was designed to predict statistical models and provide causal explanations, as Sarstedt et al. (2021) reported. PLS is a flexible statistical tool that can be applied to various types of research, including confirmatory research. This study focused on assessing the quality of

Table 3: Descriptive statistics, reliability, and convergent validity

Name	M	SD	Excess Kurtosis	Skewness	VIF	Loading	CA	Rho_A	Cr	AVE
EE1	4.781	1.818	-1.083	-0.385	3.390	0.907	0.922	0.936	0.944	0.810
EE2	4.558	1.749	-1.059	-0.226	2.939	0.908				
EE3	4.991	1.778	-0.771	-0.548	3.391	0.914				
EE4	4.969	1.794	-0.833	-0.556	2.696	0.870				
PEE1	4.397	1.934	-1.234	-0.197	2.797	0.895	0.919	0.922	0.943	0.805
PEE2	4.478	1.866	-1.040	-0.297	2.793	0.884				
PEE3	4.125	1.923	-1.144	0.018	3.655	0.912				
PEE4	4.237	1.886	-1.146	-0.028	3.364	0.897				
MOT1	4.281	1.924	-1.235	-0.094	4.339	0.944	0.937	0.940	0.960	0.888
MOT2	4.362	1.943	-1.230	-0.177	3.991	0.944				
MOT3	4.299	1.933	-1.201	-0.149	3.980	0.938				
SI1	3.344	1.830	-0.896	0.362	4.595	0.947	0.948	0.949	0.966	0.906
SI2	3.335	1.932	-0.988	0.435	4.591	0.947				
SI3	3.420	1.971	-0.991	0.413	5.747	0.961				
HAB1	4.071	1.898	-1.118	-0.056	3.382	0.921	0.925	0.926	0.952	0.869
HAB2	3.607	1.920	-1.060	0.293	3.307	0.927				
HAB3	3.982	1.892	-1.108	-0.062	4.537	0.949				

FC1	4.812	1.923	-1.149	-0.425	2.784	0.915	0.895	0.899	0.935	0.826
FC2	4.719	1.891	-0.983	-0.430	2.781	0.917				
FC3	5.214	1.782	-0.707	-0.703	2.563	0.895				
PR1	4.304	1.851	-1.057	-0.145	2.941	0.956	0.914	1172	0.941	0.841
PR2	4.393	1.827	-1.114	-0.245	3.707	0.905				
PR3	4.464	1.880	-1.121	-0.240	3.067	0.890				
TRUST1	4.388	1.898	-1.186	-0.167	3.713	0.936	0.938	0.942	0.960	0.889
TRUST2	4.446	1.913	-1.183	-0.264	4.399	0.942				
TRUST3	4.415	1.928	-1.208	-0.230	4.466	0.951				
IU1	3.879	1.925	-1.135	0.029	4.175	0.938	0.946	0.947	0.965	0.903
IU2	3.549	1.961	-1.203	0.214	4.981	0.943				
IU3	3.893	1.961	-1.206	-0.028	7.434	0.970				

Note: M = Mean; SD = Standard deviation; VIF = Variance inflation factor; CA = Cronbach's alpha; CR = Composite reliability; AVE = Average variance extracted.

fitness following the approach proposed by Henseler (2018). A non-parametric technique called bootstrapping was also incorporated, which applies to various SEM techniques, as highlighted by Henseler et al. (2016), along with a multi-group analysis to analyze the two biometric payment systems (Cheah et al., 2023).

The normality of the data was assessed using the skewness and kurtosis values of each item (Table 3). The results showed small and moderate values for both kurtosis and skewness (Curran et al., 1996). Multicollinearity was measured using variance inflation factors (VIF) The mean VIF value was 3.81, and all values were below the recommended maximum threshold of 5, except for two (SI3 and IU3). We concluded that the data do not present multicollinearity problems.

3.4. Reliability and Validity Analysis

The reliability, convergent validity, and discriminant validity of the measurement model were assessed. Four criteria were used to assess the model's reliability: factor loadings, Cronbach's alpha (CA), Rho coefficient, and composite reliability (CR) (Table 4). The item loadings for each factor were above 0.70 and significant (Barclay et al., 1995). CA, Rho coefficient, and CR values were above the recommended minimum of 0.7 (Nunnally and Bernstein, 1994). Convergent validity was measured through average variance extracted (AVE). The AVE values were above the recommended minimum of 0.5 (Fornell and Larcker, 1981). The discriminant validity of PLS is assessed using three different methods: (i) the examination of cross-loadings of indicators requires that the loadings of each indicator on its construct are greater than the cross-loadings on other constructs; (ii) the Fornell-Larcker criterion, which analyzes whether the correlations between dimensions are less than the square root of the AVE (Fornell and Larcker, 1981); and (iii) the heterotrait-monotrait (HTMT) ratio of correlations between two constructs, which should be below 0.9 (Henseler et al., 2015). Results demonstrate that the measurement model met the reliability and validity criteria.

3.5. Testing the Hypothesis Using SEM

Once the convergent and discriminant validity of the measurement model was verified, the structural model was analyzed. We tested the

Table 4: Discriminant validity

	EE	FC	HAB	IU	MOT	PEE	PR	SI	TRUST
EE	**0.900**	0.627	0.456	0.418	0.581	0.598	0.115	0.408	0.551
FC	0.696	**0.909**	0.503	0.440	0.496	0.410	0.136	0.311	0.605
HAB	0.492	0.551	**0.932**	0.752	0.618	0.662	0.047	0.668	0.638
IU	0.440	0.476	0.803	**0.950**	0.590	0.644	0.100	0.696	0.616
MOT	0.625	0.539	0.663	0.625	**0.942**	0.667	0.173	0.656	0.562
PEE	0.645	0.447	0.718	0.688	0.719	**0.897**	0.139	0.600	0.511
PR	0.126	0.147	0.068	0.094	0.186	0.132	**0.917**	0.139	-0.001
SI	0.432	0.335	0.712	0.735	0.696	0.643	0.141	**0.952**	0.526
TRUST	0.590	0.660	0.685	0.651	0.597	0.548	0.051	0.556	**0.943**

Note: The square roots of the AVEs are in bold on the main diagonal. The Fornell-Larcker criterion is depicted below the main diagonal. The heterotrait-monotrait (HTMT) ratio values were above the main diagonal.

research hypotheses through a comparative analysis of the structural coefficients. For this purpose, we performed a bootstrap analysis on 5,000 randomly drawn subsamples from the original dataset, showing that the four proposed hypotheses received empirical support. The hypotheses and corresponding results were: H1 (effort expectancy) (β EE→UI = –0.103, p-value = 0.078), H2 (performance expectancy) (β PE→UI = 0.201, p-value = 0.006), H3 (hedonic motivation) (β HM→UI = –0.013, p-value = 0.867), H4 (social influence) (β SN→UI = 0.284, p-value = 0.000), H5 (habit) (β HAB→UI = 0.342, p-value = 0.000), and H6 (facilitating conditions) (β FC→UI = 0.061, p-value = 0.372).

Finally, H7 postulated a negative effect of perceived risk on intention to use, which did not receive empirical support (β PRISK→UI = 0.022, p-value = 0.599), unlike H8, which proposed a positive relationship between trust and intention that received empirical support (β TRUST→UI = 0.173, p-value = 0.021).

Additionally, the model's predictive ability was assessed by determining the squared multiple correlation coefficient (R2). The R2 value for intention to use was 0.669, indicating that it explained a high proportion of the variance in the model. We also examined the standardized root mean square residual (SRMR) value (Henseler et al., 2015) to test the difference between the observed and predicted correlations as an indicator of the model fit. A value of less than 0.08 is considered acceptable. The model proposed in this study yielded a value below the threshold (0.046). Table 5 summarizes the results.

Table 5: Results of the hypothesis testing

Hypothesis	Relation	Path Coefficient	T-value	P-value	Supported
H1	EE → IU	-0.103	1.761	0.078	No
H2	PEE → IU	0.201	2.768	**0.006**	**Yes**
H3	MOT → IU	-0.013	0.168	0.867	No
H4	SI → IU	0.284	4.090	**0.000**	**Yes**
H5	HAB → IU	0.342	4.715	**0.000**	**Yes**
H6	FC → IU	0.061	0.893	0.372	No
H7	PR → IU	0.022	0.526	0.599	No
H8	TRUST → IU	0.173	2.307	**0.021**	**Yes**

Table 6:. Results of SEM multigroup analysis (MGA)

Hypothesis	Relation	Finger Payment		Voice Payment		Finger Payment versus Voice Payment	
		Path	P-value	Path	P-value	Difference	2-tailed P-value
H1	EE → IU	0.070	0.277	-0.109	0.440	0.179	0.242
H2	PEE → IU	0.294	0.002	0.019	0.901	0.275	0.129
H3	MOT → IU	0.038	0.681	-0.026	0.861	0.063	0.709
H4	**SI → IU**	**0.119**	**0.101**	**0.483**	**0.000**	**-0.364**	**0.020**
H5	**HAB → IU**	**0.462**	**0.000**	**0.192**	**0.093**	**0.270**	**0.050**
H6	FC → IU	-0.098	0.307	0.149	0.139	-0.248	0.071
H7	PR → IU	-0.011	0.819	-0.005	0.954	-0.005	0.943
H8	TRUST → IU	0.096	0.329	0.241	0.038	-0.145	0.338

The results show significant differences in two of the structural model relationships (p-value < 0.05): social influence and intention, and habit and intention. Specifically, the impact of social influence on intention is greater among potential users of the pay-per-voice system ($\beta_{Finger} = 0.119$, p-value = 0.101; $\beta_{Voice} = 0.483$, p-value = 0.000; $\beta_{Difference} = -0.364$, p-value = 0.020), and the impact of habit on intention is higher among potential users of fingerprint payment systems ($\beta_{Finger} = 0.462$, p-value = 0.000; $\beta_{Voice} = 0.192$, p-value = 0.093; $\beta_{Difference} = 0.270$, p-value = 0.050). No other relationships showed significant differences.

3.6. Moderating Effect of Type of MPS

PLS multigroup analysis (PLS-MGA) was used to analyze the moderating effect of MPS type, for which the sample was divided into two groups (footprint versus voice payment). In the multigroup analysis, the path coefficients of the two structural models resulting from the division of samples were compared according to the two established categories. Student's t-test for independent samples was used to assess whether significant differences existed (Ringle et al., 2015).

Prior to this analysis, invariance of the measuring instrument (MICOM) was verified to ensure that the possible differences in the model arising from the moderating effect of the variables were not caused by differences in the measurement models used for each group. If this was the case, conducting a multigroup analysis would not have been logical. The verification process confirmed configuration invariance, composite invariance, and partial measurement invariance. Therefore, a multigroup analysis was appropriate for assessing the possible moderating effect of the variables (Table 6).

4. Discussion

According to the recent survey "New habits in payment methods in Spain 2023" (PaynoPain, 2023), users' future forecasts highlight that almost 52% of Spaniards are betting on payment with biometrics (face, iris, fingerprint, or voice), followed by smartphone payments or digital wallets, which is already a reality. Given this circumstance, this study focuses on an exhaustive analysis of the factors that influence the intention to use biometric payment systems, specifically those based on fingerprint and voice recognition.

4.1. Theoretical Implications

Using the UTAUT2 model extended with the variables of perceived risk and trust, we explored how users perceive and adopt these technologies in the context of financial transactions. The results confirm that habit is the variable that has the greatest relevance to intention to use, followed by social influence, performance expectation, and trust. These results are consistent with previous studies on technology adoption (Venkatesh et al., 2012; Higueras-Castillo et al., 2023).

Although initially expected to play an important role, risk perception did not significantly impact the intention to use these technologies. This finding can be interpreted in light of the growing awareness and trust in biometric technologies, which are perceived as more secure and reliable than traditional methods (Featherman & Pavlou, 2003). Furthermore, trust emerged as a key factor in adoption, suggesting that mitigating privacy and security concerns could significantly increase users' willingness to adopt biometric payment systems (Gefen et al., 2003).

This study contributes significantly to the literature on the adoption of financial technologies by extending the UTAUT2 model by including the variables of perceived risk and trust. These two factors have been explored less in the context of biometric payment systems, but this study demonstrates their relevance in assessing consumer behavior. Confirming that trust plays a moderating role in the relationship between risk perception and intention to use also opens new avenues for investigating the impact of trust in other financial technology contexts (Hu et al., 2023).

In addition, this study delves into a comparative analysis between different types of biometrics (fingerprints and voice) and how the unique characteristics of each technology may influence adoption. The integration of Perceived Risk Theory and Trust Theory into the UTAUT2 framework provides richer insights into the psychological determinants that drive the acceptance of new technologies. This suggests that future technology adoption models should consider greater complexity in the interactions between these factors (Pavlou, 2003; Shiau et al., 2023).

Regarding the proposed moderating effect, the PLS multi-group analysis (PLS-MGA) results showed significant differences between the two types of biometric payment systems analyzed: fingerprint and voice recognition. In particular, social influence was found to have a greater impact on the intention to use voice recognition than fingerprint payment. This finding suggests that users of voice recognition-based technologies tend to be more susceptible to the influence of people in their environment, reinforcing the importance of social context in the adoption of emerging technologies (Venkatesh et al., 2012). In contrast, habit was found to be a more important determinant in the adoption of the fingerprint payment system, indicating that users who are already familiar with this type of authentication tend to use it more automatically. These results suggest that the type of biometric technology influences how psychological and social factors affect its adoption, implying that implementation strategies should be tailored to the type of technology used.

Results showed that differences between the two biometric systems were evident in the relationships between social influence and intention to use and the relationship between habit and intention to use. For voice payments, strong social influence could be related to the need for social validation in using fewer familiar technologies and the perception that their adoption aligns with emerging technology trends. However, fingerprint payments seem to benefit from smoother adoption owing to users' previous experience with this technology, which reinforces the importance of encouraging repeated use and familiarity. The difference in the moderating effects between the two technologies highlights the importance of considering the specific characteristics of each biometric system, which adds further complexity to the modified UTAUT2 model, particularly in terms of how the technologies are perceived and adopted in different environments (Cheah et al., 2023).

4.2. Managerial Implications

From a management perspective, this study offers important conclusions for stakeholders in the biometric payments market (developers, financial institutions, fintech, and paytech partner companies). First, communication strategies should focus on strengthening consumer confidence by highlighting the advanced security features of biometric systems and minimizing concerns about risk. Security is a crucial factor that, if handled properly, can enhance the adoption of these technologies (Ahmed et al., 2021).

Companies should bear in mind that social influence is a key factor in the adoption of innovative technologies. This implies that, besides conventional advertising campaigns, using public influencers and promoting recommendations from satisfied users can be effective strategies to increase visibility and trust in biometric payment systems (Belanche et al., 2022). Finally, companies must ensure that these technologies are easy to implement. Reducing barriers to entry, such as the perceived effort to learn how to use biometric systems, will result in faster and more widespread adoption, especially among users who are less familiar with the technology (Ogbanufe & Kim, 2018).

4.3. Limitations and Potential Future Research Avenues

This study had some limitations. (1) The sample used focused on a specific cultural and geographic context, which may limit the generalizability of the results to other regions. Future research should explore the adoption of these systems in different geographical and cultural contexts to validate the applicability of the findings. (2) Although this study focused on fingerprint and voice recognition, other biometric technologies merit further study, like facial recognition and iris authentication, which offer different adoption dynamics and risk perceptions (Kambampati et al., 2024).

The methodology used on a group of potential users only captures the intentions themselves without reflecting the users' own behavior, so future research should record behaviors from surveys and use other psychophysiological measures that facilitate the actual interpretation of behavior towards these financial technologies (Antiopi et al., 2024), using more complex experimental designs (Casado-Aranda et al., 2018).

Another limitation is the cross-sectional nature of the data collected, which prevents a longitudinal analysis of the adoption of biometric payment systems. It would be interesting to explore how users' perceptions evolve over time as they become more familiar with these technologies and whether trust and perceived risk change as their use increases (Juniper Research, 2021).

Finally, future research could delve deeper into demographic differences and analyze how factors like age, education level, and previous experience with technologies influence the adoption of biometric payment systems (Hamid et al., 2024).

References

Ahmed, W., Rasool, A., Javed, A.R., Kumar, N., Gadekallu, T.R., Jalil, Z., & Kryvinska, N. (2021). Security in next generation mobile payment systems: A comprehensive survey. *IEEE Access*, 9, 115932–115950.

Allied Market Research. (2022). *Biometric Payments Market Report: Global Opportunities and Forecast 2021–2026.* Retrieved from https://www.alliedmarketresearch.com.

Antiopi, P., Kalaitzi, E., & Fidas, C.A. (2024). A review on the use of EEG for the investigation of the factors that affect consumer's behavior. *Physiology & Behavior*, 1–18.

Barclay, D., Thompson, R., & Higgins, C. (1995). The Partial Least Squares (PLS) approach to causal modeling: Personal computer adoption and use as an illustration. *Technology Studies*, 2(2), 285–309.

Bauer, R.A. (1967). Consumer behavior as risk taking. *Marketing: Critical Perspectives on Business and Management*, 593, 13–21.

Belanche, D., Casaló, L.V., Flavián, C., & Schepers, J. (2022). Service robot implementation: A theoretical framework and research agenda. *The Service Industries Journal*, 40(3–4), 203–225.

Bhat, S.A., Islam, S.B., & Mir, M.F. (2024). Consumers' attitude toward biometric banking services: An empirical evaluation of determinants and outcomes. *Journal of Financial Services Marketing*, 1–17.

Blanco-Oliver, A., Lara Rubio, J., Irimia-Diéguez, A., & Liébana Cabanillas, F.J. (2024). Examining user behavior with machine learning for effective mobile peer-to-peer payment adoption. *Financial Innovation*, 10(1), 1–30.

Casado-Aranda, L.A., Liébana-Cabanillas, F., & Sánchez-Fernández, J. (2018). A neuropsychological study on how consumers process risky and secure E-payments. *Journal of Interactive Marketing*, 43(1), 151–164.

Cheah, J.H., Amaro, S., & Roldán, J.L. (2023). Multigroup analysis of more than two groups in PLS-SEM: A review, illustration, and recommendations. *Journal of Business Research*, 156, 113539.

Chukwuere, J.E., Ntseme, O.J., & Shaikh, A.A. (2021). Toward the development of a revised technology acceptance model. In: *Proceedings of the International Conference on Electronic Business. International Consortium for Electronic Business*. December 3-7, 2021. Nanjing, China.

Curran, P.J., West, S.G., & Finch, J.F. (1996). The robustness of test statistics to nonnormality and specification error in confirmatory factor analysis. *Psychological Methods*, 1(1), 16.

European Commission. (2023). Payment Services Directive (PSD3) – Revised rules on payment services and electronic money. Available at https://ec.europa.eu.

Faster Capital. (2024). *The Benefits of Biometric Payments*. FasterCapital. https://fastercapital.com/keyword/the-benefits-of-biometric-payments.html

Featherman, M.S., & Pavlou, P.A. (2003). Predicting e-services adoption: A perceived risk facets perspective. *International Journal of Human-Computer Studies*, 59(4), 451–474.

Fornell, C., & Larcker, D.F. (1981). Evaluating structural equation models with unobservable variables and measurement error. *Journal of Marketing Research*, 18(1), 39–50.

Gefen, D., Karahanna, E., & Straub, D.W. (2003). Trust and TAM in online shopping: An integrated model. *MIS Quarterly*, 27(1), 51–90.

Gomez-Barrero, M., Maiorana, E., Galbally, J., Campisi, P., & Fierrez, J. (2017). Multi-biometric template protection based on homomorphic encryption. *Pattern Recognition*, 67, 149–163.

Guan, Y., Wang, D., Chu, Z., Wang, S., Ni, F., Song, R. et al. (2023). *Intelligent Virtual Assistants with LLM-based Process Automation* (No. arXiv:2312.06677). arXiv. https://doi.org/10.48550/arXiv.2312.06677

Hamid, H.R.M.H., Nordin, N.W., Abdullah, N.Y., Ismail, W.H.W., & Abdullah, D. (2024). Two factor authentication: voice biometric and token-based authentication. In: *Applied Problems Solved by Information Technology and Software* (pp. 27–35). Springer Nature Switzerland.

Henseler, J. (2018). Partial least squares path modeling: Quo vadis? *Quality & Quantity*, 52(1), 1–8.

Henseler, J., Ringle, C.M., & Sarstedt, M. (2015). A new criterion for assessing discriminant validity in variance-based structural equation modeling. *Journal of the Academy of Marketing Science*, 43, 115–135.

Henseler, J., Sinkovics, R.R., Ringle, C.M., & Sarstedt, M. (2016). Testing measurement invariance of composites using partial least squares. *International Marketing Review*, 33(3), 405–431.

Hu, B., Liu, Y.L., & Yan, W. (2023). Should I scan my face? The influence of perceived value and trust on Chinese users' intention to use facial recognition payment. *Telematics and Informatics*, 78, 101951.

Juniper Research. (2021). *Biometric Payments: Adoption Challenges & Opportunities 2020–2025*. Available at https://www.juniperresearch.com/press/biometrics-to-secure-over-3-trillion-in-mobile/#

Kambampati, P., Rane, S., Shoeb, A., & Dhannawat, R. (2024, July). PAYV-Payment Voice: A platform using voice recognition to enable payment transactions. In: *2024 Asia Pacific Conference on Innovation in Technology (APCIT)* (pp. 1–6). IEEE.

Kathuria, R., Wadehra, A., & Kathuria, V. (2020). Human-centered artificial intelligence: Antecedents of trust for the usage of voice biometrics for driving contactless interactions. In: *HCI International 2020 – Late Breaking Posters: 22nd International Conference*, Copenhagen, Denmark, July 19–24, 2020, Proceedings, Part I 22 (pp. 325–334). Springer International Publishing.

Kim, C., Mirusmonov, M., & Lee, I. (2010). An empirical examination of factors influencing the intention to use mobile payment. *Computers in Human Behavior*, 26(3), 310–322.

Kumar, T. R., Babu, D. V., Malarvezhi, P., Velu, C. M., Haritha, D., & Karthikeyan, C. (2021). Boltzmann–Dirichlet process mixture: A mathematical model for speech recognition. *Journal of Physics: Conference Series*, 1964(4), 042039. https://doi.org/10.1088/1742-6596/1964/4/042039

Kumar, S. (2024). Biometric systems security and privacy issues. In: *Leveraging Computer Vision to Biometric Applications*. Chapman and Hall/CRC.

Kumari, K. (2024). Improving payment security with deep learning-based facial recognition systems in mobile banking applications. *Journal of Sustainable Technologies and Infrastructure Planning*, 8(3), 13–20.

Lee, C., & Pan, L.-Y. (2023). Resistance of facial recognition payment service: A mixed method approach. *Journal of Services Marketing*, 37, 392-407. https://doi.org/10.1108/JSM-01-2022-0035

Liébana-Cabanillas, F., Ramos de Luna, I., & Montoro-Ríos, F. (2017). Intention to use new mobile payment systems: A comparative analysis of SMS and

NFC payments. *Economic Research* (*Ekonomska istraživanja*), 30(1), 892–910.

Liébana-Cabanillas, F., Muñoz-Leiva, F., & Sánchez-Fernández, J. (2018). A global approach to the analysis of user behavior in mobile payment systems in the new electronic environment. *Service Business*, 12, 25–64.

Liébana-Cabanillas, F., Japutra, A., Molinillo, S., Singh, N., & Sinha, N. (2020). Assessment of mobile technology use in the emerging market: Analyzing intention to use m-payment services in India. *Telecommunications Policy*, 44(9), 102009.

Liébana-Cabanillas, F., Singh, N., Kalinic, Z., & Carvajal-Trujillo, E. (2021). Examining the determinants of continuance intention to use and the moderating effect of gender and age of users of NFC mobile payments: A multi-analytical approach. *Information Technology and Management*, 22(2), 133–161.

Liébana-Cabanillas, F., Kalinić, Z., Muñoz-Leiva, F., & Higueras-Castillo, E. (2023). Biometric m-payment systems: A multi-analytical approach to determining use intention. *Information & Management*, 61, 103907. https://doi.org/10.1016/j.im.2023.103907

Ling, P.S., Lim, X.J., Wong, L.J., & Lee, K.Y.M. (2024). Thank you, government! Your support facilitated my intention to use mobile payment in the new normal era. *Spanish Journal of Marketing-ESIC*, 29(1), 3–21.

Madan, K., & Yadav, R. (2017). Understanding and predicting antecedents of mobile shopping adoption: A developing country perspective. *Asia Pacific Journal of Marketing and Logistics*, 30. https://doi.org/10.1108/APJML-02-2017-0023

McKnight, D.H., Choudhury, V., & Kacmar, C. (2002). Developing and validating trust measures for e-commerce: An integrative typology. *Information Systems Research*, 13(3), 334–359.

Migliore, G., Wagner, R., Cechella, F.S., & Liébana-Cabanillas, F. (2022). Antecedents to the adoption of mobile payment in China and Italy: An integration of UTAUT2 and innovation resistance theory. *Information Systems Frontiers*, 24(6), 2099–2122. https://doi.org/10.1007/s10796-021-10237-2

Molinillo, S., & Higueras-Castillo, E. (2022). Do biometric payment systems work during the COVID-19 pandemic? Insights from the stimulus-organism-response model. *Financial Innovatio*n, 8(1), 1–22.

Nguyen, T.D., & Le, N.T. (2022). Determinants of intention to use FinTech payment services. *International Journal of Business, Economics and Law*, 26(1), 354–366.

Nunnally, J.C., & Bernstein, I.H. (1994). *Psychometric Theory*. New York: McGraw-Hill Companies. https://books.google.es/books?id=r0fuAAAAMAAJ

Ogbanufe, O., & Kim, D.J. (2018). Comparing fingerprint-based biometrics authentication versus traditional authentication methods for e-payment. *Decision Support Systems*, 106, 1–14.

Oliveira, T., Thomas, M., Baptista, G., & Campos, F. (2016). Mobile payment: Understanding the determinants of customer adoption and intention to recommend the technology. *Computers in Human Behavior*, 61, 404–414. https://doi.org/10.1016/j.chb.2016.03.030

Pavlou, P.A. (2003). Consumer acceptance of electronic commerce: Integrating trust and risk with the technology acceptance model. *International Journal of Electronic Commerce*, 7(3), 101–134.

PaynoPain. (2023, October). *New Habits in Payment Methods in Spain 2023.* PaynoPain. https://paynopain.com/en/blog/new-payment-habits-in-spain-2023-the-card-and-smartphone-era/

Ringle, C.M., Wende, S., & Becker, J.M. (2015). SmartPLS 3. SmartPLS GmbH, Boenningstedt. *Journal of Service Science and Management*, 10(3), 32–49.

Sarstedt, M., Ringle, C.M., & Hair, J.F. (2021). Partial least squares structural equation modeling. *Handbook of Market Research*, 26, 1–40.

Seng, W., & Hee, O. (2021). Factors influencing the intention to use E-wallet: An extended hedonic-motivation system adoption model. *International Journal of Academic Research in Business and Social Sciences*, 11. https://doi.org/10.6007/IJARBSS/v11-i11/11379

Shiau, W.L., Liu, C., Zhou, M., & Yuan, Y. (2023). Insights into customers' psychological mechanism in facial recognition payment in offline contactless services: Integrating belief–attitude–intention and TOE–I frameworks. *Internet Research*, 33(1), 344–387.

Singh, S., Singh, N., Kalinić, Z., & Liébana-Cabanillas, F.J. (2021). Assessing determinants influencing continued use of live streaming services: An extended perceived value theory of streaming addiction. *Expert Systems with Applications*, 168, 114241.

Slade, E.L., Dwivedi, Y.K., Piercy, N.C., & Williams, M.D. (2015). Modeling consumers' adoption intentions of remote mobile payments in the United Kingdom: Extending UTAUT with innovativeness, risk, and trust. *Psychology & Marketing*, 32(8), 860–873. https://doi.org/10.1002/mar.20823

Sarstedt, M., Ringle, C., & Hair, J. (2021). *Partial Least Squares Structural Equation Modeling* (pp. 1–47). https://doi.org/10.1007/978-3-319-05542-8_15-2

Soto-Beltrán, L., Robayo, O., & Rojas-Berrio, S. (2020). Effects of perceived risk on intention to use biometrics in financial products: Evidence from a developing country. *International Journal of Business Information Systems*, 1, 1. https://doi.org/10.1504/IJBIS.2020.10028631

Statista. (2024). *Mobile OS Market Share Worldwide 2009-2024.* Market Share of Mobile Operating Systems Worldwide from 2009 to 2024, by Quarter. https://www.statista.com/statistics/272698/global-market-share-held-by-mobile-operating-systems-since-2009/

Taiwo, A.A., & Downe, A.G. (2013). The theory of user acceptance and use of technology (UTAUT): A meta-analytic review of empirical findings. *Journal of Theoretical and Applied Information Technology*, 49(1), 48-58.

Venkatesh, V., Morris, M.G., Davis, G.B., & Davis, F.D. (2003). User acceptance of information technology: Toward a unified view. *MIS Quarterly*, 27(3), 425–478. https://doi.org/10.2307/30036540

Wadood, A., Ali, Z., Ghouzali, S., ALfawaz, B., Muhammad, G., & Hossain, M.S. (2017). Biometric security through visual encryption for fog edge computing. *IEEE Access*, 1–1. https://doi.org/10.1109/ACCESS.2017.2693438

Williams, M.D., Rana, N.P., & Dwivedi, Y.K. (2015). The unified theory of acceptance and use of technology (UTAUT): A literature review. *Journal of Enterprise Information Management*, 28(3), 443–488.

Yadav, R., & Sharma, R. (2023). Factors Influencing the adoption intention of using mobile financial services during the COVID-19 pandemic: The role of FinTech. *Environmental Science and Pollution Research*, 30, 61271–61289.

Venkatesh, V., Thong, J.Y., & Xu, X. (2012). Consumer acceptance and use of information technology: Extending the unified theory of acceptance and use of technology. *MIS Quarterly*, 36(1), 157–178.

Wang, J.S. (2021). Exploring biometric identification in FinTech applications based on the modified TAM. *Financial Innovation*, 7(1), 42. https://doi.org/10.1186/s40854-021-00260-2

Wells, W.D. (Ed.). (2014). *Measuring Advertising Effectiveness*. Psychology Press. https://doi.org/10.4324/9781315806099

8

Integration of Blockchain and Biometrics: Applications, Benefits, and Challenges in Key Sectors

Juan F. Prados-Castillo[1*(0000-0003-3588-5534)], Eva R. Porras[2(0000-0002-2167-6275)], Konstantinos Sgantzos[3(0000-0001-9033-5805)] and Francisco Liébana-Cabanillas[1(0000-0002-3255-0651)]

[1] University of Granada
[2] Universidad Rey Juan Carlos
[3] National Technical University of Athens

1. Introduction

The integration of blockchain and biometric technology has been explored in several studies, particularly in terms of improving reliability, security, and efficiency. Some publications (Shukla & Bhushan, 2023; Dalal et al., 2020) have proposed the use of biometric data to access and verify the information stored in the blockchain, namely, the management of educational certifications and personal identities. Malek et al. (2022) and Delgado-Mohatar (2020) highlight the potential of Blockchain to protect and manage sensitive biometric data, especially in aviation security. However, the combination of these technologies presents challenges, like latency, processing time, and protection of sensitive personal data (Delgado-Mohatar et al., 2019; Ghafourian et al., 2023). Despite these challenges, the potential benefits of their integration make them a promising area for future research and development for

*Corresponding author: jfprados@ugr.es

enterprise applications (Ahmad et al., 2022; Singh et al., 2022). Such integration can provide enhanced reliability, security, and privacy for various applications in the real economy, including biometric technologies, like fingerprints, iris scanning, or facial recognition that can be used to authenticate users and protect access to blockchain-based systems. Blockchain technology, with its decentralized and tamper-resistant nature, can be leveraged to store and manage biometric data securely. This union can be particularly beneficial in areas like financial transactions, health records, and identity management, where both strong authentication and secure data storage are crucial (Halkiopoulos et al., 2023).

An example of this integration is the use of a multimodal biometric authentication system (those that combine several biometric systems) that seeks to leverage machine learning and blockchain technology. In this approach, multiple biometric modalities, namely, fingerprints, faces, and irises, are merged and stored in a blockchain network, providing a higher level of security and trust in user authentication. One important aspect is the ability of the blockchain to pseudonymize these data by hashing the biometric record for each user and storing it in the blockchain. This adds value to user privacy and creates a secure environment upon which the user's personal information is secured. The immutable and distributed nature of the blockchain ensures the integrity and privacy of biometric data, whereas machine learning algorithms improve the accuracy and reliability of the authentication process (Sgantzos & Grigg, 2019). This combination helps address the computational cost and complexity associated with signing and verifying transactions in blockchain networks (Sooksomsatarn et al., 2021). The need for complex cryptographic operations can be reduced by leveraging biometric authentication, which results in efficient and cost-effective blockchain transactions.

The aim of this work is to explore and analyze the applications, benefits, and challenges of integrating biometrics and blockchain technology. It aims to highlight the potential of the mixed technology to improve security, privacy, and efficiency in sectors like banking, identity management, and medical records. Based on this argument, the following research questions are:

RQ1: What are the benefits of integrating blockchain and biometrics in terms of security and privacy?

RQ2: How can Blockchain protect and manage sensitive biometric data?

RQ3: What are the most promising applications of blockchain and biometrics integration?

RQ4: What technical challenges does the integration of blockchain and biometrics face?

RQ5: How can multimodal biometric authentication improve efficiency and security in blockchain?

RQ6: Which sectors can benefit the most from the integration of these technologies?

2. Fundamentals of Biometrics

2.1. Biometrics: Definition, Characteristics, and Comparative Advantages

Biometric technology is defined as the automated identification of individuals based on their unique physiological or behavioral characteristics (Jaiswal et al., 1970; Gaur et al., 2012). The most common physical biometric characteristics are fingerprints, hand geometry, iris, retina, and facial patterns, while behavioral biometrics encompass voice, signature, and gait (Aloudat et al., 2016; Barbole & Godase, 2023). Table 1 presents some of the commonly used biometric technologies.

Biometric authentication offers advantages over traditional methods, like passwords or tokens, because biometric traits cannot be easily lost, stolen, or forged (Lott, 2018). This technology has evolved from its early applications in high-security facilities to its widespread use in various sectors, including finance, access control, and identity management (Lawton, 1998; Asha & Chellappan, 2012). Biometric systems typically involve processes for capturing data through sensors, processing and comparing them with previously stored templates for verification or identification (Manasi, 2015).

As the demand for increased security increases, the biometric market is expected to expand significantly. However, there are challenges related to accuracy, privacy, and interoperability (Lawton, 1998). Unimodal biometric systems, characterized by using a single biometric trait for authentication, face several limitations, namely, noisy data, lack of universality, impersonation attacks, and unacceptable error rates. In turn, whereas multimodal biometric systems, characterized by using two or more biometric traits, have emerged as a solution to overcome the limitations of unimodal systems (Sondhi & Bansal, 2014).

Table 1: Main biometric technologies and their sectoral applications

Technology	Utility	Benefits	Challenges	Authors
Vascular Biometry	Secure authentication using vascular patterns	High accuracy, difficult to forge, effective implementation (e.g., Japan since 2004)	Sensitive to environmental conditions (light, temperature)	Sanchez-Reillo et al., 2009; Sarkar et al., 2016
Fingerprints	-	Accessible to rural users, higher accuracy compared to passwords and PINs	-	Ahamed et al., 2022; Palaniappan et al., 2010; Das & Debbarma, 2011
Iris Scanning	Secure authentication for ATMs and online transactions	High accuracy, difficult to counterfeit, fast authentication, combines with other biometrics for greater reliability	Integration with existing infrastructure; handling of data compression algorithms	Al-Debei et al., 2024; Omolara et al., 2019
Facial Recognition	-	Frictionless, accurate experience even with changes in appearance or illumination	-	Sudhakar et al., 2024
Voice Recognition	Secure authentication for ATMs and mobile banking	Convenient, natural authentication during conversation	Cost of implementation, integration challenges with other technologies	Singh et al., 2022

Retinal Identification	-	Maximum security in sensitive environments	-	Mubeen, 2024
Vein Recognition	Identification and authorization of banking transactions	Difficult to counterfeit, high accuracy	Sensitive to privacy, possible biases in algorithms	Boudjellal, 2024; Mubeen, 2024
DNA Identification	-	Highest accuracy in critical transactions	-	Hassan & Sabahat, 2024; Mohammed0 et al., 2024
Behavioral Biometrics	Authentication in phone banking services and mobile applications	No additional hardware required, easy integration into digital platforms	Sensitive to environmental noise, voice sickness and recording attacks	Elhag & Alharbi, 2024; C. Wang et al., 2024

Source: Author's own elaboration.

Evolution and Applications of Biometric Systems

From a management point of view, biometric systems have been applied in sectors, like the banking sector, using vascular biometrics (Sarkar et al., 2010), which uses unique patterns of veins in the palm, eye, or ocular surface. These systems are very accurate because the vascular patterns are located internally in the body and are extremely difficult to spoof. The palm vein authentication system has been implemented in banks in Japan since 2004, showing reliable and highly accurate results in high-traffic contexts (Sanchez-Reillo et al., 2009; Qamber et al., 2012).

Another widely adopted biometric is fingerprint recognition, which provides secure authentication at ATMs and online transaction platforms. This is particularly useful in rural areas, where access to advanced technologies is limited, and helps make the system accessible and secure for more users (Palaniappan et al., 2010; Ahamed et al., 2022).

Iris scanning represents another promising banking technology, noted for its high security. This technique is difficult to spoof and is suitable for implementation in ATMs and mobile applications. Iris recognition offers a high degree of accuracy and reliability, which are crucial factors in banking authentication and is comparatively superior to other biometric methods owing to its stability and uniqueness (Vincy & Sathana, 2019). It has been proven effective in preventing bank fraud and helping to reduce incidents like phishing and spoofing, which aim at obtaining sensitive user information through deception (Omolara et al., 2019).

How Biometric Technology Works and Challenges

The implementation of biometrics in business practice faces certain limitations, namely problems caused by environmental factors and privacy concerns. Environmental conditions, like changes in light or temperature, can affect the performance of vascular biometric systems, making them difficult to use in environments exposed to extreme variations in temperature or light (Sanchez-Reillo et al., 2009). The collection and storage of biometric data raises privacy concerns regarding the security of user information. To address these challenges, current research has focused on the development of techniques to protect stored biometric templates (Hartung, 2012). Additionally, the costs associated with the infrastructure to implement biometric systems can be high, representing a financial challenge for some institutions (Maček et al., 2019). To mitigate these challenges, hashing and storing the biometric record for each user on the blockchain can be utilized. This procedure can be done easily by the employment of a smart contract, and the record of the user will be represented by a unique number that cannot be reversed engineered to point to the users' identity.

The safekeeping of biometric records requires robust hashing techniques that ensure data security and privacy while allowing certain functionalities, like authentication or comparison. Some of the key characteristics of an adequate hash for biometric records are:

1. **Irreversibility (One-way hashing):** The hash function must make it computationally infeasible to reverse-engineer the original biometric data from the hash.
2. **Uniqueness (Collision resistance):** The hash function must ensure that two different biometric inputs produce distinct hash outputs.
3. **Consistency (Deterministic):** Given the same biometric input, the hash function should always produce the same output.
4. **Tolerance for Variance (Fuzzy hashing):** Biometric data often vary slightly with each capture (e.g., changes in lighting or angle for fingerprints). The hashing process should allow approximate matching without storing the original data.
5. **Salted Hashing:** To prevent attackers from using precomputed hash tables (e.g., rainbow tables), each biometric hash should incorporate a unique 'salt' value.

The steps below reproduce a secure workflow example:

1. Enrollment Phase:
   - Obtain biometric data (e.g., fingerprint and iris scans).
   - Normalize and transform it to reduce variability.
   - Generate a unique salt for this purpose
   - Apply the hash function to the biometric data and salt.
   - The hash and salt are stored in the database.
2. Authentication Phase:
   - Obtain biometric data for authentication.
   - Normalize and transform these data using the same procedure as in enrollment.
   - The transformed data were combined with the stored salt.
   - Hash the combined input and compare it with the stored hash for verification.

Some other considerations include:

1. **Compliance:** Ensure adherence to regulations such as GDPR (Regulation - 2016/679 - EN - GDPR - EUR-Lex, 2016), which require privacy-preserving measures for biometric data storage.

2. **Tamper-proof Storage:** Hashed biometric records are stored in a tamper-proof environment, such as a blockchain or a secure hardware module.
3. **Regular Updates:** Periodically update or review the salting and hashing mechanisms to ensure their validity with respect to advances in computational power or attacks.

Using these techniques, biometric records can be securely stored and processed, ensuring privacy and security against modern threats.

3. Blockchain Technology

Recently, Blockchain technology has garnered increasing scholarly attention. For instance, according to Forbes, the global cryptocurrency market cap as of 27 November, 2024, stands at $3.46 Trillion (https://www.forbes.com/digital-assets/crypto-prices) as compared to $1.67T by the end of January 2022 (Coinmarketcap, 2022). Furthermore, this technology elicits substantial expectations from global investment forecasts. By 2030, the impact on global economic growth is projected to be US$1.76 trillion (PwC, 2020). From a practical perspective, the technology demonstrates significant potential to reduce financial costs, secure transactions, facilitate contract dispute resolution processes, and develop more reliable and transparent loyalty systems (Prados-Castillo et al., 2023[1]). This is due to the attributes of a technology that can transform sectors plagued by inefficiencies and ease the start of new business models (Porras-Gonzalez et al.[2])

Four key characteristics determine the blockchain's potential to become the seed of numerous processes.

1. *Decentralization of certification*, which replaces a central intermediary.
2. *Fast execution of activities* given that exchanges happen directly between the parties to the transaction.
3. *Immediate accessibility*, given that the only infrastructure required is Internet access, no one can prevent someone else's access to this tool.

---

[1] A Critical Analysis Of The Advantages Brought By Blockchain Technology To The Global Economy. *Int. J. Intellectual Property Management*, 9(2), (co-Authors Martín-Martín, J.M. and Guaita-Martínez, J.M.) June 2019. https://doi.org/10.1504/IJIPM.2019.100214.

[2] Same as above (Footnote 1).

4. *Security:* Blockchain technology is protected by asymmetric cryptography, an economic model of self-interest, and other modern and sophisticated mechanisms.

However, other qualities of this technology are also central to its uses. These are:

1. *Confidentiality* that restricts access to information in the data.
2. *Integrity* results from ensuring the consistency and accuracy of the data over its lifespan.
3. *Authentication* occurs by the automatic verification of the identity of a user and the ownership of the tokens.
4. *Non-repudiation* which ascertains the account owner stays liable for the actions of the account (Menezes et al., 1996[3]).

The listed characteristics, explain why the blockchain is expected to be the foundation of an Industrial Revolution as it replaces legacy technologies with one that is "independent of boundaries and intermediaries, cheaper, faster, and more secure to implement" (Porras-Gonzalez et al. p. 9[4]). Blockchain is considered a general-purpose technology (Böhme et al., 2015; Tapscott and Tapscott, 2016a[5]) with many potential applications (Porras-Gonzalez et al.[6]).

Since the publication of the seminal paper on Bitcoin (Nakamoto, 2008), the development of Blockchain technology has progressed to its application in the 'Crypto Economy'. This term denotes a transaction system that utilizes cryptographic hashes as computational proof mechanisms within a network, with the objective of transferring and confirming the transmission of assets, such as electronic cash or smart contracts. Blockchain technology aims to supplement the trust traditionally placed in third-party intermediaries with a series of tools

3 Menezes, A.J., Van Oorschot, P.C. and Vanstone, S.A. (1996). Information security and cryptography. In: *Handbook of Applied Cryptography* (Chapter 1.2), pp. 2–4 [online] https://web.archive.org/web/20050305140333/http://www.cacr.math.uwaterloo.ca:80/hac/about/chap1.pdf (accessed 25 November 2018).

4 Same as Footnote 1.

5 Böhme, R., Christin, N., Edelman, B. and Moore, T. (2015). Bitcoin: Economics, technology, and governance. *The Journal of Economic Perspectives*, 29(2), pp. 213–238.

Tapscott, D. and Tapscott, A. (2016a). *Blockchain Revolution: How the Technology Behind Bitcoin is Changing Money, Business, and the World*, Penguin Canada.

Tapscott, D. and Tapscott, A. (2016b). The impact of the blockchain goes beyond financial services. *Harvard Business Review* [online] https://hbr.org/2016/05/the-impact-of-theblockchain-goes-beyond-financial-services?autocomplete=true (accessed 27 December 2018).

6 Same as Footnote 1.

that render these organizations superfluous. It substitutes trust in human actions with a mix of processes and technologies, including algorithms based on blockchain technology, attributing trust characteristics to the economic system it underpins (Swan, 2015). From a practical perspective, blockchain technology demonstrates potential value for generating business and social value in the corporate sphere due to its innovative features in terms of security, transparency, and privacy management (Deloitte, 2016). However, this is also due to its speed of transaction completion and cost reduction.

Numerous practical applications have been implemented in recent years based on the utilization of blockchain technology through the development of Smart Contracts (Table 2) (Rashideh, 2020). These contracts can automatically execute a set of instructions specified in a computerized protocol (Szabo, 1995). For instance, a Smart Contract could execute the cancellation of a tourist accommodation booking contract if the end customer has not confirmed the credit card details that guarantee the transaction.

Table 2: Applications and utilities of blockchain technology

Technology	Utility
Reducing financial costs	Reduction of financial costs in corporate transactions and processes.
Secure transactions	Implementation in financial systems and contractual dispute resolution.
Reliable and transparent loyalty systems	Development of loyalty systems in tourism and other industries.
Crypto-economics and algorithmic trust	Replacement of traditional financial institutions with Blockchain algorithms.
Smart contracts	Automatic execution of protocols for reservations and transactions.
Innovation in security and transparency	Creation of social and business value through secure and transparent data management.

Source: Author's own elaboration.

4. Synergy between Biometrics and Blockchain

The integration of blockchain technology and biometrics technology offers promising opportunities to improve security, privacy, and efficiency in various sectors (Ghafourian et al., 2023; Mykola & Andriy, 2024). This synergy can improve processes related to identity management and authentication systems (Delgado-Mohatar et al., 2019;

Sharma et al., 2024). Blockchain's immutability and decentralization complement the uniqueness and reliability of biometric technologies, potentially revolutionizing areas such as aviation security and civil identity management (Malek et al., 2022; Shukla & Bhushan, 2023).

However, potential challenges remain, namely, some blockchain's limitations in terms of processing capacity and storage demands (Mykola & Andriy, 2024). This calls for careful scrutiny of the specific blockchain chosen for these processes, given that each has its own protocol and defined capacities. Economic, legal, and ethical issues, like costs, the allocation of responsibilities, and proper data protection assessments need attention (Ghafourian et al., 2023). Despite these obstacles, ongoing research is exploring innovative applications, namely, the incorporation of biometric attributes within blockchain hash functions and the development of smart contracts for biometric data management (Delgado-Mohatar et al., 2019; Dinesh et al., 2021). Figure 1 shows the space generated between the field of blockchain technology and biometric technology.

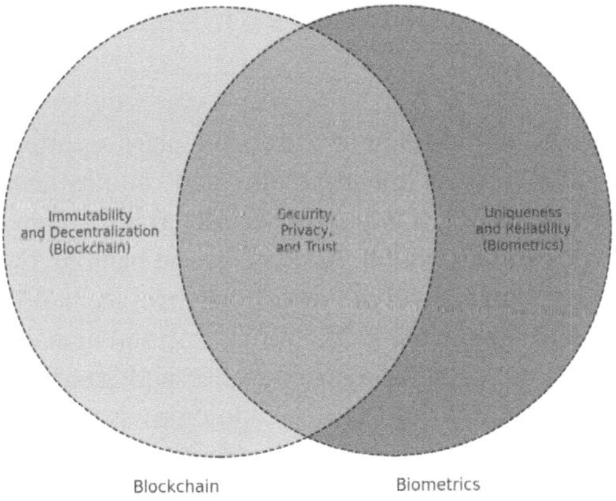

Figure 1: Synergy between blockchain and biometrics. *Source:* Own elaboration.

The integration of blockchain and biometrics seeks to provide greater security, privacy, and efficiency in identity management (Mykola & Andriy, 2024). This synergy can improve the protection of stored biometric data and enable novel applications, including distributed trust services (Shukla & Bhushan, 2023). Hence, despite these challenges, blockchain-based biometric systems offer promising solutions for secure and privacy-preserving identity management, allowing individuals greater control over their biometric data (Liu et al., 2019). Future developments can focus on improving the efficiency and scalability while maintaining the security benefits of this technology integration.

Blockchain enables secure user authentication and payments in financial transactions (Baralla et al., 2019; Sharma et al., 2024). Medical records ensure secure management of medical data and controlled access to patient information (Barka et al., 2022). Identity management facilitates digital identification and protects against identity theft (Shakya, 2019; Paez et al., 2020). Integration improves the security and efficiency of voting systems (Department of Computer Science and Engineering, 2024), increases the security of IoT networks (Hassen et al., 2020), and enables the outsourcing of biometric identification to preserve privacy (Banu & Hani, 2022). Regardless of the challenges-in balancing security and scalability (Hassen et al., 2020), the integration of biometrics and blockchain holds promise for strengthening authentication, preserving privacy, and improving access control in various industries.

5. Applications of Biometrics-Blockchain Integration

The integration of blockchain and biometrics offers promising applications in identity management and authentication systems (Ghafourian et al., 2023). It improves security, transparency, and privacy in various sectors, like aviation, healthcare, and banking (Mudliar et al., 2018). From a technical point of view, blockchain technology provides a decentralized and immutable ledger for storing and managing biometric data, addressing concerns of data breaches and unauthorized access. Although the integration of these technologies is still in preliminary development, it shows the potential to improve the security and efficiency of biometric systems, although legal and accountability issues require further consideration. Four specific applications for the integration of these technologies are developed in the following sections.

5.1. Health Registrations

The integration of biometrics and blockchain technology into healthcare systems has the ability to provide enhanced security, privacy, and access control to electronic health records (EHRs) (Al Baqari & Barka, 2020; Barka et al., 2022). This application addresses challenges, namely, data integrity, patient identification, and emergency access without compromising privacy. Blockchain-based systems can securely store and share patient data collected through IoT devices and biometric systems (Gautam et al., 2023). For example, the integration of global identity systems, as used by India (UIDAI, 2024), with blockchain technology would enable the development of a variety of healthcare-related applications (Mudliar et al., 2018; Khatoon et al., 2019). These systems have the technical capability to grant decentralized, secure, and transparent access to complete medical records across treatment centers (Dhagarra et al., 2019). Overall, the combination of biometrics and blockchain in healthcare offers enhanced data security, patient privacy, and efficient access to medical information (Kumar & Agrawal, 2024). Figure 2 shows an example of an operation for health record protection that integrates both technologies.

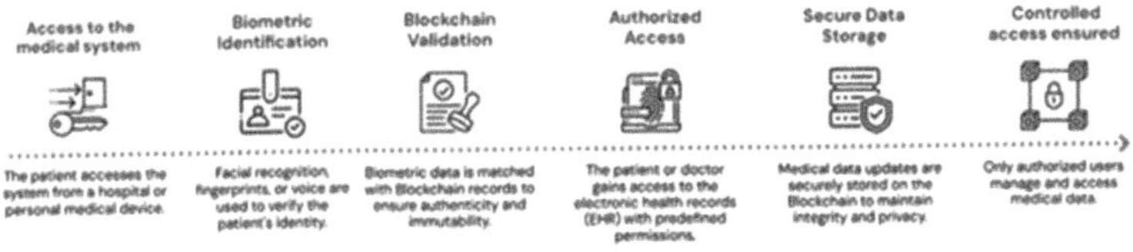

Figure 2: Healthcare records management flow chain integrating blockchain and biometrics. *Source:* Own elaboration from @Freepik designs.

5.2. Financial Transactions

The integration of blockchain and biometrics offers promising solutions to improve security in financial transactions (Mykola & Andriy, 2024). This synergy can improve authentication processes, protect previously stored biometric templates, and ensure data integrity (Delgado-Mohatar et al., 2019, 2020). Blockchain's immutability and decentralization complement the uniqueness of biometric recognition, potentially revolutionizing digital identity systems and financial services (Shukla & Bhushan, 2023). Advanced applications include multi-factor authentication using facial recognition and voice verification for secure transactions (Laila & Tamilpavai, 2023). However, some challenges remain, like blockchain's limited transaction processing capacity, storage requirements, and legal concerns regarding accountability and data protection. However, the combination of blockchain and biometrics shows potential to create robust, efficient, and secure systems in various sectors, namely, banking and cloud computing (Manoj, 2023).

Figure 3 shows an example of a system operation for securing financial transactions that integrates both technologies analyzed.

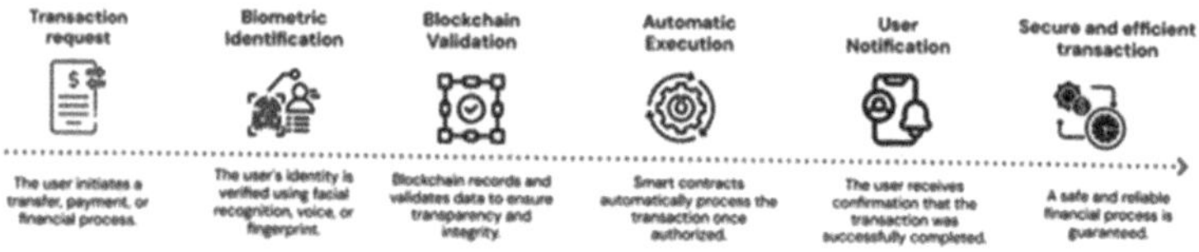

Figure 3: Security flow of financial transactions with blockchain and biometrics. *Source:* Own elaboration from @Freepik designs.

5.3. Identity Management

Blockchain technology offers decentralized and tamper-proof storage for biometric data, thereby addressing privacy issues associated with centralized systems (Gao et al., 2018; Liu et al., 2019). Smart contracts enable flexible access control policies, allowing users to manage their biometric data securely (Jacob et al., 2021). Blockchain-based systems have demonstrated high performance in facial recognition tasks with low error rates reported for popular datasets (Salem et al., 2020). Several studies have proposed key management mechanisms and protection schemes for previously stored biometric templates to improve security and privacy (Odelu, 2019). A promising solution is to hash the user's biometric data and store the hash on the blockchain. In this way, the user will be the only one who knows the identity and the hash, while the hash record on the blockchain will remain anonymous. The integration of blockchain, self-sufficient identity, and facial recognition technology is a promising approach to streamline authentication processes (Pawade et al., 2019). Efficient biometric-based anonymous credential schemes have been developed for smart industrial applications, offering features such as selective disclosure and revocation while ensuring compliance with data protection regulations and legislation (Sarier, 2021). Figure 4 shows an example of a digital identification system in which both technologies were analyzed.

5.4. Loyalty and Customer Engagement

Blockchain technology offers promising solutions for improving customer loyalty programs in various industries. It addresses traditional bottlenecks, like coupon loss and redemption complications (Bulbul

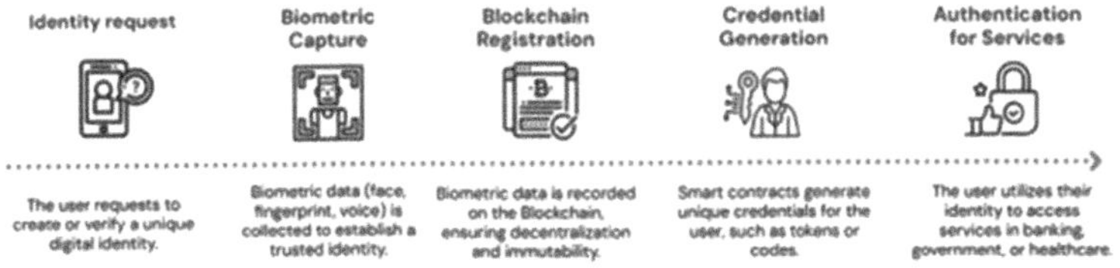

Figure 4: Digital identity management flow with blockchain and biometrics. *Source:* Own elaboration from @Freepik designs.

& Ince, 2018), while providing enhanced security, privacy, and data integrity (Kumar & Agrawal, 2024). Blockchain-based loyalty systems also have the ability to reduce administrative costs, improve customer experiences, and increase user engagement (Udegbe, 2017). The integration of biometrics with blockchain presents opportunities to improve the protection of customer data stored with biometric data (Delgado-Mohatar et al., 2019). The integration of these technologies could include increasing security in advanced point exchange systems between different loyalty programs (Nguyen et al., 2023) and creating more secure loyalty systems (Santos et al., 2023). Figure 5 shows an example of a system for improving loyalty and customer engagement analyzed using both technologies.

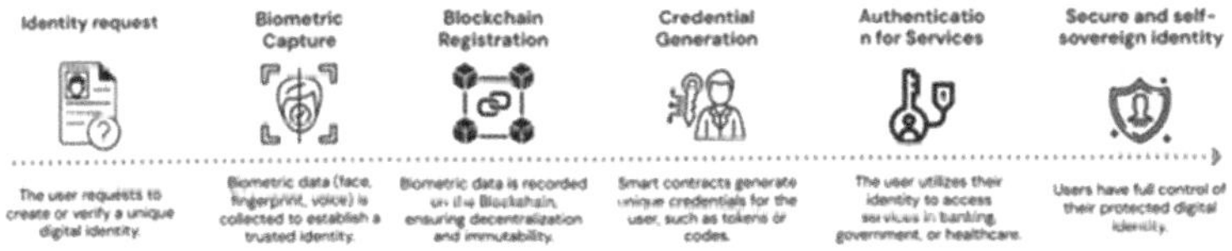

Figure 5: Loyalty and customer engagement flow based on blockchain and biometrics. *Source:* Own elaboration from @Freepik designs.

6. Opportunities and Challenges in Blockchain-Biometrics Integration

The integration of blockchain and biometrics presents opportunities and challenges. Blockchain technology can improve security, privacy, and decentralization in biometric systems (Ghafourian et al., 2023). This integration offers the possibility to develop applications that include protection of previously stored biometric templates, identity management, and authentication (Sharma et al., 2024). The implementation of blockchain for biometric solutions faces technical obstacles, namely, latency, processing time, and economic costs (Delgado-Mohatar et al., 2020). Legal and regulatory challenges include liability allocation and compliance with data protection (Yeoh, 2017). However, the integration of blockchain and biometrics holds promise for improving financial inclusion and self-reliant identity systems (Wang & De Filippi, 2020). As the field evolves, it will be crucial to address technical, operational, and legal challenges to realize the full potential of blockchain and biometrics integration (Lopes et al., 2019).

7. Challenges and Strategic Recommendations for Implementation

The implementation of this technology in critical sectors aims to improve security in financial transactions, medical records, and identity systems, minimizing fraud risks and increasing reliability (Halkiopoulos et al., 2023). By enabling secure storage of biometric data, blockchain facilitates users with greater control over their personal data, contributing to increased adoption in areas where privacy is paramount. As authentication systems evolve to become more secure, this integration is expected to drive innovations in IoT network security, voting, and distributed trust services, although these advances need to be balanced with scalability and associated costs. Table 3 provides guidance on recommendations for future implementation.

Although integration of blockchain and biometrics encounters significant challenges, its potential benefits in terms of enhanced security and privacy present a compelling area for future research and practical applications. This technological convergence offers a potential pathway towards a future in which authentication systems exhibit increased security, trustworthiness, and user-centricity. However, the effective implementation of such systems will be contingent upon advancements in the technical, legal, and social adoption domains.

8. Conclusions

This study has several limitations. First, since it is based on the analysis of existing documents, it is possible that more recent or relevant studies have been omitted due to restrictions in the selection of sources, time coverage, or availability in certain databases. Furthermore, its exploratory

Table 3: Recommendations for the Implementation and Development of Systems Based on Blockchain and Biometric Data

Functionality	Description	Industry	Authors
Improvement of technical infrastructure	Invest in advanced techniques to reduce latency and improve storage of biometric templates.	Financial, Health, Technology	Delgado-Mohatar et al., 2019
Development of specific regulations	Establish a clear legal framework on privacy, responsibilities and user rights.	Legal, Government, Corporate	Ghafourian et al., 2023
Focus on user education	Improve understanding of benefits and limitations to reduce privacy fears.	Education, Marketing, Health	(Antoniadis et al., 2020; Kumar & Agrawal, 2024)N/A
Exploration of scalability solutions	Consider alternatives such as sidechaining and holomorphic encryption to improve efficiency.	Technology, Health, Research	Sharma & Dwivedi, 2024

Source: Author's own elaboration.

nature implies that the findings should be interpreted as a starting point for future research rather than as definitive conclusions. The need for complementary studies that delve deeper into the issues addressed and incorporate primary data and more exhaustive analyses is underscored.

The combination of blockchain and biometrics technology offers a highly secure and tamper-resistant authentication system. Biometrics, by its unique and inherent nature to each individual, ensures accurate identification, whereas blockchain's decentralization and immutability provide secure storage, preventing tampering and unauthorized access to sensitive data (Dalal et al., 2020; Shukla & Bhushan, 2023). This integration faces technical challenges, like latency and processing times, which affect system efficiency. Concerns about data protection, high infrastructure costs, and legal and ethical constraints on user privacy also persist (Singh et al., 2022).

Proposals like the hashing of personal user biometric data could add value and mitigate concerns about data protection, while several blockchains can utilize methods such as Elliptic Curve Digital Signature Algorithm (ECDSA) and hashing to safely encrypt and ensure that the data is owned and can only be accessed by their respectful owners. Specifically, for Bitcoin, these are Secp256k1 and Secure Hash Algorithm 256 (SHA256) (Nakamoto, 2008). It must be noted that Machine Learning over encrypted data demonstrates that it is feasible to perform computations directly on encrypted datasets, which can be personal medical data, without decrypting them first (Dwork, 2006). This capability is crucial for applications involving biometric data analysis because it allows organizations to derive valuable insights while preserving user privacy and security.

It is also essential to address the potential challenges associated with implementing such systems:

- **Key Management:** Cryptographic key security is critical. Weak key management can significantly undermine the integrity of an entire system.
- **Scalability:** Biometric data hashing must maintain both algorithmic efficiency and robust security as data volumes increase.
- **Regulatory Compliance:** Privacy-preserving techniques must precisely align with data protection regulations such as GDPR to ensure legal and ethical data handling.

Other hashing techniques and procedures are often recommended for handling biometric data.

- **Fuzzy Hashing or Secure Sketch Techniques:** Standard cryptographic hashes like SHA-256 may sometimes not be directly suitable for biometrics because they require identical inputs to produce the same output. In these instances, other methods that permit small variations in the input to still produce matching results are used. Some examples are fuzzy hashing methods, such as Locality-Sensitive Hashing (LSH) or biometric-specific secure sketches.
- **Cancelable Biometrics:** By transforming the biometric data into a reversible but secure format before hashing and creating a transformed version of the biometric data, an extra layer of protection is added when combined with a hashing process.
- **Salting:** This process is strengthened by combining the biometric data with a random, unique value (the salt) before hashing. Salt should be stored in the hash to allow verification.

- **Using Strong Cryptographic Hashes:** For the actual hashing process, strong cryptographic algorithms such as SHA-256 or SHA-3 are recommended. These are collision-resistant and are widely regarded as secure for general cryptographic purposes.
- **Encryption of Hash Values:** Encrypting hash values adds an additional layer of security. This ensures that even if the hashes are leaked, they remain computationally difficult to use without the encryption key.
- **Zero-knowledge Proofs for Authentication:** For biometric verification, zero-knowledge protocols can be used to prove that a captured biometric matches a stored hash without revealing the biometric or the hash itself.

References

Hassen, O.A., Abdulhussein, A.A., Darwish, S.M., Othman, Z.A., Tiun, S., & Lotfy, Y. A. (2020). Towards a secure signature scheme based on multimodal biometric technology: Application for IOT blockchain network. *Symmetry*, 12(10), 1699. Crossref. https://doi.org/10.3390/sym12101699.

Ahamed, J., Maisha, M., Labiba, Z., Islam, Md. A., & Nandi, D. (2022). A review report on the fingerprint-based biometric system in ATM banking. In: *Proceedings of the 2nd International Conference on Computing Advancements*, 522–529. https://doi.org/10.1145/3542954.3543029.

Ahmad, I., Khusro, S., Alam, I., Khan, I., & Niazi, B. (2022). Towards a low-cost teacher orchestration using ubiquitous computing devices for detecting student's engagement. *Wireless Communications & Mobile Computing*, 2022, 7979766. https://doi.org/10.1155/2022/7979766.

Al Baqari, M., & Barka, E. (2020, June). Biometric-based Blockchain EHR System (BBEHR). In: *2020 International Wireless Communications and Mobile Computing (IWCMC)*. Crossref. https://doi.org/10.1109/iwcmc48107.2020.9148357.

Al-Debei, M.M., Hujran, O., & Al-Adwan, A.S. (2024). Net valence analysis of iris recognition technology-based FinTech. *Financial Innovation*, 10(1), 59. https://doi.org/10.1186/s40854-023-00509-y.

Aloudat, A., Michael, K., & Abbas, R. (2016). The implications of iris-recognition technologies: Will our eyes be our keys? *IEEE Consumer Electronics Magazine*, 5(3), 95–102. Crossref. https://doi.org/10.1109/mce.2016.2556901.

Antoniadis, I., Spinthiropoulos, K., & Kontsas, S. (2020). Blockchain applications in tourism and tourism marketing: A short review. In: A.

Kavoura, E. Kefallonitis, & P. Theodoridis (Eds.), *Strategic Innovative Marketing and Tourism* (pp. 375–384). Springer International Publishing. https://doi.org/10.1007/978-3-030-36126-6_41.

Asha, S., & Chellappan, C. (2012). Biometrics: An overview of the technology, issues and applications. *International Journal of Computer Applications*, 39(10), 35–52. Crossref. https://doi.org/10.5120/4859-7134.

Banu, Dr. S., & E Hani, B.H.U. (2022). Mobile fingerprint verification and automatic log in platform using blockchain. *International Journal for Research in Applied Science and Engineering Technology*, 10(11), 737–741. Crossref. https://doi.org/10.22214/ijraset.2022.47256.

Baralla, G., Ibba, S., Marchesi, M., Tonelli, R., & Missineo, S. (2019). A blockchain based system to ensure transparency and reliability in food supply chain. Lecture Notes in Computer Science (Including Subseries Lecture Notes in Artificial Intelligence and Lecture Notes in Bioinformatics), 11339 *LNCS*, 379–391. https://doi.org/10.1007/978-3-030-10549-5_30/FIGURES/2.

Barbole, A., & Godase, M. (2023). Advancements in biometrics provides additional security. *SSRN Electronic Journal*. Crossref. https://doi.org/10.2139/ssrn.4607940.

Barka, E., Al Baqari, M., Kerrache, C.A., & Herrera-Tapia, J. (2022). Implementation of a biometric-based blockchain system for preserving privacy, security, and access control in healthcare records. *Journal of Sensor and Actuator Networks*, 11(4), 85. Crossref. https://doi.org/10.3390/jsan11040085.

Boudjellal, S.E. (2024). *Biometric Security System* [PhD Thesis]. http://dspace.univ-setif.dz:8888/jspui/handle/123456789/4371.

Bulbul, S., & Ince, G. (2018). Blockchain-based framework for customer loyalty program. In: *2018 3rd International Conference on Computer Science and Engineering (UBMK)*, 342–346. Crossref. https://doi.org/10.1109/ubmk.2018.8566642.

Coinmarketcap. (2022). *Precios, gráficos y capitalizaciones de mercado de criptomonedas | CoinMarketCap*. https://coinmarketcap.com/es/.

Dalal, J., Chaturvedi, M., Gandre, H., & Thombare, S. (2020). *Verification of Identity and Educational Certificates of Students Using Biometric and Blockchain* (SSRN Scholarly Paper No. 3564638). Social Science Research Network. https://doi.org/10.2139/ssrn.3564638

Das, S.S., & Debbarma, Smt. J. (2011). Designing a biometric strategy (fingerprint) measure for enhancing ATM security in Indian e-banking system. *International Journal of Information and Communication Technology Research*, 1(5), 7. https://citeseerx.ist.psu.edu/document?repid=rep1&type=pdf&doi=0d9ff1280accdb90ee959fe02efe527b72ec375e

Delgado-Mohatar, O., Fierrez, J., Tolosana, R., & Vera-Rodriguez, R. (2019). Biometric Template storage with blockchain: A first look into cost and performance tradeoffs. In: *2019 IEEE/CVF Conference on Computer Vision and Pattern Recognition Workshops* (*CVPRW*), 2829–2837. Crossref. https://doi.org/10.1109/cvprw.2019.00342.

Delgado-Mohatar, O., Fierrez, J., Tolosana, R., & Vera-Rodriguez, R. (2020). *Blockchain meets Biometrics: Concepts, Application to Template Protection, and Trends*. https://doi.org/10.48550/arXiv.2003.09262

Deloitte. (2016). *Beyond Bitcoin: Blockchain is Coming to Disrupt Your Industry*. https://www2.deloitte.com/mt/en/pages/financial-services/articles/mt-banking-alert-019-blockchain-is-coming-to-disrupt-your-industry.html.

Department of Computer Science and Engineering, Jawaharlal Nehru New College of Engineering. (2024). Biological authentication voting system using blockchain. *International Journal of Scientific Research in Engineering and Management*. https://doi.org/10.55041/ijsrem34142.

Dhagarra, D., Goswami, M., Sarma, P.R.S., & Choudhury, A. (2019). Big Data and blockchain supported conceptual model for enhanced healthcare coverage. *Business Process Management Journal*. Crossref. https://doi.org/10.1108/bpmj-06-2018-0164.

Dinesh, A.D., Durga Prasad Reddy, C., Gopi, G.V., Jain, R., & Shankar, T.N. (2021). A durable biometric authentication scheme via blockchain. In: *2021 International Conference on Advances in Electrical, Computing, Communication and Sustainable Technologies* (*ICAECT*), 1–5. Crossref. https://doi.org/10.1109/icaect49130.2021.9392415.

Dwork, C. (2006). Differential privacy. In: M. Bugliesi, B. Preneel, V. Sassone, & I. Wegener (Eds.), *Automata, Languages, and Programming* (Vol. 4052, pp. 1–12). Springer Berlin Heidelberg. https://doi.org/10.1007/11787006_1.

Elhag, S., & Alharbi, R.S.S. (2024). Behavioral biometrics: A case study of digital banking security in Saudi Arabia. In: C. Hewage, L. Nawaf, & N. Kesswani (Eds.), *AI Applications in Cyber Security and Communication Networks* (pp. 39 51). Springer Nature. https://doi.org/10.1007/978-981-97-3973-8_3.

Gao, Z., Xu, L., Turner, G., Patel, B., Diallo, N., Chen, L., & Shi, W. (2018, June). Blockchain-based identity management with mobile device. In: *Proceedings of the 1st Workshop on Cryptocurrencies and Blockchains for Distributed Systems*. Crossref. https://doi.org/10.1145/3211933.3211945.

Gaur, S., Shah, V., & Thakker, M. (2012). Biometric recognition techniques: A review. *International Journal of Advanced Research in Electrical, Electronics, and Instrumentation Energy*. https://www.semanticscholar.org/paper/Biometric-Recognition-Techniques%3A-A-Review-Gaur-Shah/ce247f03c636f9c0505c71a0bb47ad82734841a5.

Gautam, K.K., Prakash, S., & Dwivedi, R.K. (2023). Patients medical record monitoring using IoT based biometrics blockchain security system. *2023 International Conference on IoT, Communication, and Automation Technology* (*ICICAT*), 1–6. Crossref. https://doi.org/10.1109/icicat57735.2023.10263595.

Ghafourian, M., Sumer, B., Vera-Rodriguez, R., Fierrez, J., Tolosana, R., Moralez, A., & Kindt, E. (2023). *Combining Blockchain and Biometrics: A Survey on Technical Aspects and a First Legal Analysis*. https://doi.org/10.48550/ARXIV.2302.10883.

Halkiopoulos, C., Antonopoulou, H., & Kostopoulos, N. (2023). Utilizing blockchain technology in various applications to secure data flows. A comprehensive analysis. *Technium: Romanian Journal of Applied Sciences and Technology*, 11, 44–55. https://doi.org/10.47577/technium.v11i.9132.

Hartung, D. (2012). Vascular Pattern Recognition: And its Application in Privacy-Preserving Biometric Online-Banking Systems [Doctoral thesis]. 244. https://ntnuopen.ntnu.no/ntnu-xmlui/handle/11250/144366.

Hassan, W., & Sabahat, N. (2024). Towards secure identification: A comparative analysis of biometric authentication techniques. *VFAST Transactions on Software Engineering*, 12(1). https://doi.org/10.21015/vtse.v12i1.1745.

Jacob, A.B., Prakash, P., Karhana, P., & Swain, P. (2021). ChainAccess: Blockchain based web-access through biometrics. In: *2021 12th International Conference on Computing Communication and Networking Technologies (ICCCNT)*, 1–6. Crossref. https://doi.org/10.1109/icccnt51525.2021.9579481.

Jaiswal, S., Bhadauria, D.S.S., & Jadon, D.R.S. (1970). Biometric: Case study. *Journal of Global Research in Computer Sciences*, 2(10), 19–49. https://www.rroij.com/open-access/biometric-case-study-19-49.pdf

Khatoon, A., Verma, P., Southernwood, J., Massey, B., & Corcoran, P. (2019). Blockchain in energy efficiency: Potential applications and benefits. *Energies*, 12(17), 3317. https://doi.org/10.3390/en12173317.

Kumar, R., & Agrawal, A. (2024). Using blockchain to secure biometric healthcare apps. *Biometric Technology Today*, 2024(4). Crossref. https://doi.org/10.12968/s0969-4765(24)70028-9.

Laila, P.N.J., & Tamilpavai, D.G. (2023). FinBTech: Blockchain-based video and voice authentication system for enhanced security in financial transactions utilizing FaceNet512 and Gaussian Mixture models. arXiv.Org. https://doi.org/10.48550/ARXIV.2310.18668.

Lawton, G. (1998). Biometrics: A new era in security. *Computer*, 31(8), 16–18. Crossref. https://doi.org/10.1109/mc.1998.707612.

Liu, Y., Sun, G., & Schuckers, S. (2019, June). Enabling secure and privacy preserving identity management via smart contract. In: *2019 IEEE Conference on Communications and Network Security (CNS)*. Crossref. https://doi.org/10.1109/cns.2019.8802771.

Lopes, A.C., Rodrigues, E.B., & Vera-Cruz, R. (2019). Tourism gentrification. In: *3rd World Multidisciplinary Civil Engineering, Architecture, Urban Planning Symposium (WMCAUS 2018)* (Vol. 471). IOP Publishing Ltd. https://doi.org/10.1088/1757-899X/471/9/092025.

Lott, D. (2018). Biometrics: Modernising customer authentication for financial services and payments. *Journal of Payments Strategy & Systems*, 12(4), 371. Crossref. https://doi.org/10.69554/cgak3521.

Maček, N., Adamović, S., Milosavljević, M., Jovanovic, M., Gnjatović, M., & Trenkić, B. (2019). Mobile banking authentication based on cryptographically secured iris biometrics. *Acta Polytechnica Hungarica*, 16, 45–62.

Malek, O.O., Islam, Z.O., Ahmed, M.U.O., Matin, M., Dasari, V. R., & Hall, R.D. (2022, May 30). *Fusion of Blockchain and Biometrics: An Integrated Platform for Aviation Security and Authentication*. https://doi.org/10.1117/12.2624966

Manasi, V. (2015). A study of biometrics technology methods and their applications: A review. *Computer Science*. http://ijiet.com.

Manoj, K.S. (2023). Secure blockchain banking cloud with error recovery processes. In: *2023 3rd International Conference on Innovative Practices in Technology and Management* (*ICIPTM*), 1–6. Crossref. https://doi.org/10.1109/iciptm57143.2023.10118028.

Mohammed, S.M., Mashhadani, S., & Hassen, O.A. (2024). Human biometric identification: Application and evaluation. *International Journal of Engineering in Computer Science*, 6(2), 131–152. https://doi.org/10.33545/26633582.2024.v6.i2b.136.

Mubeen, M. (2024). Biometric authentication: Past, present, and future perspectives. *International Journal of Innovative Research in Technology and Science*, 12(2).

Mudliar, K., Parekh, H., & Bhavathankar, P. (2018). A comprehensive integration of national identity with blockchain technology. In: *2018 International Conference on Communication Information and Computing Technology* (*ICCICT*), 1–6. Crossref. https://doi.org/10.1109/iccict.2018.8325891.

Mykola, K., & Andriy, K. (2024). Blockchain and biometrics challenges and solutions. *Computer Design Systems Theory and Practice*. https://doi.org/10.23939/cds2024.01.189.

Nakamoto, S. (2008). *Bitcoin: A Peer-to-Peer Electronic Cash System*. https://bitcoin.org/bitcoin.pdf.

Nguyen, L.-T., Nguyen, D.-T., Ngoc, K.N.N., & Duc, D.T.V. (2023). Blockchain adoption in logistics companies in Ho Chi Minh City, Vietnam. *Cogent Business and Management*, 10(2). Scopus. https://doi.org/10.1080/23311975.2023.2216436.

Odelu, V. (2019). IMBUA: Identity management on blockchain for biometrics-based user authentication. In: *Advances in Intelligent Systems and Computing* (pp. 1–10). Springer International Publishing; Crossref. https://doi.org/10.1007/978-3-030-23813-1_1.

Omolara, A.E., Jantan, A., Abiodun, O.I., Arshad, H., & Mohamed, N.A. (2019). Fingereye: Improvising security and optimizing ATM transaction time based on iris-scan authentication. *International Journal of Electrical and Computer Engineering* (*IJECE*), 9(3), 1879. Crossref. https://doi.org/10.11591/ijece.v9i3.pp1879-1886.

Palaniappan, An. R., Swaminathan, R., Rajkumar, S.B., & Sandhya, M.K. (2010). Bio-cashier: Using Bezier curve data compression and advanced encryption. *ICoAC 2010*, 70–73. Crossref. https://doi.org/10.1109/icoac.2010.5725364.

Pawade, D., Sakhapara, A., Andrade, M., Badgujar, A., & Adepu, D. (2019). Implementation of Fingerprint-based authentication system using blockchain. In: *Advances in Intelligent Systems and Computing* (pp. 233–242). Springer Singapore; Crossref. https://doi.org/10.1007/978-981-13-3600-3_22.

Prados-Castillo, J.F., Guaita Martínez, J.M., Zielińska, A., & Gorgues Comas, D. (2023). A review of blockchain technology adoption in the tourism industry from a sustainability perspective. *Journal of Theoretical and Applied Electronic Commerce Research*, 18(2). https://doi.org/10.3390/jtaer18020042.

PwC. (2020). *Blockchain Technologies Could Boost the Global Economy US$1.76 Trillion by 2030*. https://www.pwc.com/gx/en/news-room/press-releases/2020/blockchain-boost-global-economy-track-trace-trust.html.

Qamber, S., Waheed, Z., & Akram, M.U. (2012). Personal identification system based on vascular pattern of human retina. *2012 Cairo International Biomedical Engineering Conference (CIBEC)*, 64–67. Crossref. https://doi.org/10.1109/cibec.2012.6473297.

Rashideh, W. (2020). Blockchain technology framework: Current and future perspectives for the tourism industry. *Tourism Management*, 80, 104125. https://doi.org/10.1016/j.tourman.2020.104125.

Regulation - 2016/679 - EN - GDPR - EUR-Lex (2016). https://eur-lex.europa.eu/eli/reg/2016/679/oj.

Salem, S., Cooper, J., Schneider, J., Croft, H., & Munro, I. (2020). Student acceptance of using augmented reality applications for learning in pharmacy: A pilot study. *Pharmacy*, 8(3), 122. https://doi.org/10.3390/pharmacy8030122.

Sanchez-Reillo, R., Fernandez-Saavedra, B., Liu-Jimenez, J., & Kwon, Y.-B. (2009). Changes to vascular biometric system security & performance. *IEEE Aerospace and Electronic Systems Magazine*, 24(6), 4–14. Crossref. https://doi.org/10.1109/maes.2009.5161717.

Santos, A.F., Marinho, J., & Bernardino, J. (2023). Blockchain-based loyalty management system. *Future Internet*, 15(5), 161. Crossref. https://doi.org/10.3390/fi15050161.

Sarier, N.D. (2021). Efficient biometric-based identity management on the blockchain for smart industrial applications. *Pervasive and Mobile Computing*, 71, 101322. Crossref. https://doi.org/10.1016/j.pmcj.2020.101322.

Sarkar, I., Alisherov, F., Kim, T., & Bhattacharyya, D. (2010). Palm vein authentication system: A review. *International Journal of Control and Automation*, 3(1), 8. https://www.earticle.net/Article/A147507.

Sarkar, A., Abbott, A.L., & Doerzaph, Z. (2016). Biometric authentication using photoplethysmography signals. In: *2016 IEEE 8th International Conference on Biometrics Theory, Applications and Systems* (*BTAS*), 1–7. https://doi.org/10.1109/BTAS.2016.7791193.

Sgantzos, K., & Grigg, I. (2019). Artificial intelligence implementations on the blockchain. Use cases and future applications. *Future Internet*, 11(8). https://doi.org/10.3390/fi11080170.

Shakya, S. (2019). Efficient security and privacy mechanism for block chain application. *Journal of Information Technology and Digital World*, 01(02), 58–67. Crossref. https://doi.org/10.36548/jitdw.2019.2.001

Sharma, S., Saini, A., & Chaudhury, S. (2024). Multimodal biometric user authentication using improved decentralized fuzzy vault scheme based on Blockchain network. *Journal of Information Security and Applications*, 82, 103740. Crossref. https://doi.org/10.1016/j.jisa.2024.103740.

Shukla, H., & Bhushan, B. (2023). Empowering biometrics authentication system using decentralized blockchain based applications. In: *2023 International Conference on Computing, Communication, and Intelligent Systems* (*ICCCIS*), 1177–1182. Crossref. https://doi.org/10.1109/icccis60361.2023.10425327.

Singh, A., Dhanaraj, R.K., Ali, Md. A., Balusamy, B., & Sharma, V. (2022). Blockchain technology in biometric database system. In: *2022 3rd International Conference on Computation, Automation, and Knowledge Management* (*ICCAKM*), 1–6. https://doi.org/10.1109/ICCAKM54721.2022.9990133.

Sondhi, K., & Bansal, Y. (2014). *Concept of Unimodal and Multimodal Biometric System*. https://www.semanticscholar.org/paper/Concept-of-Unimodal-and-Multimodal-Biometric-System-Sondhi-Bansal/2eccf8f401b739b7a80301eb721ebb6aad07c548.

Sooksomsatarn, K., Et Al., & Et Al. (2021). Computational cost reduction of transaction signing in blockchain. *Psychology and Education Journal*, 58(1). https://doi.org/10.17762/pae.v58i1.935.

Sudhakar, K., Raju, N.P.K., Praveen, V., & Hithin, M. (2024). Advanced authentication system: A secure authentication methodology incorporating face and text passwords, image password grids, sensitive information retrieval, and login activity tracking with efficient database connection. *International Journal of Research in Engineering, Science, and Management*, 7(2).

Swan, M. (2015). *Blockchain: Blueprint for a New Economy* (O'Reilly Media, Ed.).

Szabo, N. (1995). *Smart Contracts Glossary*. https://www.fon.hum.uva.nl/rob/Courses/InformationInSpeech/CDROM/Literature/LOTwinterschool2006/szabo.best.vwh.net/smart_contracts_2.html.

Udegbe, S.E. (2017). Impact of blockchain technology in enhancing customer loyalty programs in airline business. *International Journal of Innovative Research and Advanced Studies*, 4(6). https://www.ijiras.com/2017/Vol_4-Issue_6/paper_44.pdf

UIDAI. (2024). Unique Identification Authority of India. https://www.uidai.gov.in/en/

Vincy, A.D., & Sathana, S. (2019). Recognition technique for ATM based on iris technology. *International Journal of Engineering Research & Technology*, 7(11). https://doi.org/10.17577/IJERTCONV7IS11094.

Wang, C., Tang, H., Zhu, H., Zheng, J., & Jiang, C. (2024). Behavioral authentication for security and safety. *Security and Safety*, 3, 2024003. https://doi.org/10.1051/sands/2024003.

Wang, F., & De Filippi, P. (2020). Self-sovereign identity in a globalized world: Credentials-based identity systems as a driver for economic inclusion. *Frontiers in Blockchain, 2*. Crossref. https://doi.org/10.3389/fbloc.2019.00028.

Yeoh, P. (2017). Regulatory issues in blockchain technology. *Journal of Financial Regulation and Compliance*, 25(2), 196–208. Crossref. https://doi.org/10.1108/jfrc-08-2016-0068.

Index

I

M

P

Q

R

S

T

U

V

For Product Safety Concerns and Information please contact our EU representative GPSR@taylorandfrancis.com Taylor & Francis Verlag GmbH, Kaufingerstraße 24, 80331 München, Germany